"The Pseudo Wigner Distribution for the Analysis of Doppler Ultrasound Signals"

Thesis dissertation presented by
Dario Fresa

Thesis directed by Prof. Eng. Felice Cennamo.

Submitted to the University of Naples "FEDERICO II" - Italy

February 28th 1995

"Laurea" Degree in Electronic Engineering

UNIVERSITA' DEGLI STUDI DI NAPOLI "FEDERICO II"

FACOLTA' DI INGEGNERIA

CORSO DI LAUREA IN INGEGNERIA ELETTRONICA

TESI DI LAUREA

LA PSEUDO DISTRIBUZIONE DI WIGNER NELL'ANALISI DEI SEGNALI ECOGRAFICI

Relatore:
Ch.mo Prof.
Ing. FELICE CENNAMO

Candidato:
DARIO FRESA
Matr. 15 -15049

ANNO ACCADEMICO 1993-94

DEDICATION

To My Whole Family

ACKNOWLEDGMENTS

I would like to express my deep gratefulness to Prof. Eng. Felice Cennamo who always was patient during the thesis preparation no matter how busy he was, and always had time to talk and discuss with passion about this work. I am thankful also for his friendship and willingness to talk about anything at anytime.

Finally I'd like to thanks all the Colleagues and the Researchers of the Electronic Measures Lab where I've spent really great time working to this Thesis.

ABSTRACT

"The Pseudo Wigner Distribution for the Analysis of Doppler Ultrasound Signals"
by Dario Fresa
1995

The time-frequency representations of the signals are a powerful tool for the analysis of non-stationary signals such as the biological ultrasound Doppler pulsed type. The measure of their instantaneous frequencies and of the amplitude of their spectral components, which corresponds to the measure of moving tissue velocity, can be performed using two specific types of time-frequency representation: the spectrogram, which is a well-established method of representation and of analysis, and the pseudo Wigner distribution that exhibits better temporal and frequency resolution properties, although the presence of interference terms limits its performance. The two representations, implemented in MATLAB-SIMULINK environment using optimized algorithms which have been listed in the Appendix, are compared analysing a signal simulated from a simple statistical model of Doppler ultrasound echoes coherently demodulated. Comparing the average of the standard deviation of the estimated error of the speed for the two representations, when varying the normalized bandwidth of the Doppler signal and assuming an high value of signal to noise ratio, we can note that the pseudo Wigner distribution exhibit an higher performance than the spectrogram, in agreement with the theoretical properties and with the visual analysis of images. Therefore, the analysis of Doppler ultrasound signals, for medium to high values of signal to noise ratio, can be improved replacing the spectrogram with the pseudo Wigner distribution. The performance of the latter, however, is drastically reduced for low values of signal to noise ratio, due to the presence of interference terms. This negative characteristic leads to investigate other smoothed Wigner distributions, with a greater capacity of noise suppression and for which the smoothing properties can be adjusted according to the characteristics of the signal to be analyzed.

TABLE OF CONTENTS

CHAPTER 4

Analysis of the pulsed doppler signals by mean of the spectrogram and of the pseudo Wigner distribution.

Introduction

The classic method adopted to represent the information frequencial content of a stationary signal involves the use of its Fourier transform (FT) or spectrum. For a continuous signal x (t), the FT is defined as:

$$X(f) = F(x(t)) = \int_{-\infty}^{+\infty} x(t)e^{-j2\pi ft}dt.$$

We note that the information, within the frequency domain, is defined upon an infinite time support and that, for a non-stationary signal, the temporal variation of the frequency information is not detectable. In many applications, such as those in the field of radar, sonar, acoustics, speech analysis, geophysics, biology, biomedicine and so on, the assumption of stationarity of the signal is not valid. Therefore, for time-varying signals, their analysis, representation, processing and synthesis can be performed using a time-frequency representation.

The ultrasonic echoes emitted from moving tissues are non-stationary signals and for this reason their characteristics, such as the Doppler spectrum, vary during the time. Some of the methods that, starting in the early 60s, have been adopted for the extraction of information frequencial content from the ultrasound Doppler signals, are those in the time domain, the spectral analysis, and more recently the time-frequency representations methods. Among these time-frequency representations we are going to take in consideration the spectrogram and the pseudo Wigner distribution, performing a comparison of their performances.

CHAPTER 1

NOTES ABOUT THE LINEAR AND QUADRATIC TIME-FREQUENCY

REPRESENTATIONS OF SIGNALS.

1.1 INTRODUCTION

The time-frequency representations (TFR) feature an one-dimensional signal x(t), depending on the time variable, by a two-dimensional time and frequency function $T_x(t, f)$, so, combining the time-domain analysis with the frequency domain analysis, they allow us an easy time localization of the different spectral components on the time-frequency plane. An effective analogy can be made between TFR and a musical score which indicates the notes (spectral components) that appear in every moment of the song as well as between the song and the signal to be analyzed.

The TFR has been applied successfully to analyze, edit, and synthesize non-stationary signals or time varying. They allow us, by mean of a signal processor, to identify which spectral components of a signal or of a system change during the time, allowing, at the same time, the possibility of a graphical analysis of the signal through the three-dimensional diagrams of its surfaces. The inversion algorithms have been used for the synthesis of a signal starting from a model of TFR, for the time-varying filtering, the noise suppression, an efficient signal encoding, signal detection, parameter estimation, and so on.

1.2 THE REPRESENTATION OF SIGNALS IN THE TIME AND FREQUENCY DOMAIN.

The Fourier transform (FT) and its inverse (IFT):

$$X(f) = F(x(t)) = \int_{-\infty}^{+\infty} x(t)e^{-j2\pi ft}dt \qquad (1.1)$$

$$x(t) = IF(X(f)) = \int_{-\infty}^{+\infty} X(f)e^{j2\pi ft}df \qquad (1.2),$$

establish a relationship between the representation of the signal x(t) in time domain and the frequency domain, using the spectrum X(f). The two representations are alternative ways to characterize a signal. Although the FT allows the passage from one domain to another, it shows the limitation of not providing their combination. Most of the time information is not easily accessible in the frequency domain, in fact while the spectrum X(f) shows which frequency and with which intensity they are contained in a signal, it does not provide direct information about their temporal location. This can be seen, but often with difficulties of interpretation, from the phase spectrum $\arg\{X(f)\}$. The function of *instantaneous frequency* (IF):

$$f_x(t) = \frac{1}{2\pi} \frac{d}{dt} \arg\{x(t)\} \qquad (1.3),$$

where x(t) is a complex signal (if x(t) is real one we consider its analytic version) and the function *group delay* (GD):

$$t_x(f) = -\frac{1}{2\pi} \frac{d}{df} \arg\{X(f)\} \qquad (1.4),$$

are able to describe the temporal localization of spectral components but only for a very limited class of signals [1].

1.3 THE TIME-FREQUENCY REPRESENTATIONS

The limitations associated with the FT, IF and GD can be overcome by describing the structure time-frequency of a signal through a designated surface on the time-frequency plane. From a mathematical point of view, this corresponds to a joint function $T_x(t, f)$ of time t and frequency f that we can define *time-frequency representation* of the signal x(t). The Appendix A provides definitions and properties of the main TFR. The type of relation between the specific TFR and the signal that can be linear, quadratic or nonlinear represents a key element for their classification.

1.4 THE TIME-FREQUENCY LINEAR REPRESENTATIONS

All the linear TFR satisfies the superposition principle or linearity principle according to which, if x(t) is a linear combination of certain signals, its TFR is a linear combination of the TFR of the same signals:

$$x(t) = ay(t) + bz(t) \Rightarrow aT_y(t, f) + bT_z(t, f) \qquad (1.5).$$

Linearity is a very useful characteristic in applications which involve multi-component signals. The linear TFR of basic importance are the short Fourier transform (STFT) and wavelet transforms or wavelets (WT).

1.4.1 THE SHORT FOURIER TRANSFORM

Starting from the need for a localization temporal tool of the frequency components comes the definition of *Short Fourier transform* (STFT) or "*short spectrum*" of a signal x(t):

$$STFT_X^\gamma (t, f) = \int_{t'} \left[x(t') \gamma^*(t'-t) \right] e^{-j2\pi f t'} dt' \qquad (1.6).$$

It is clear that the STFT, evaluated at time t, is the FT of the signal x(t') multiplied by an "analysis window" $\gamma^*(t'-t)$ shifted around instant t. Since the multiplication by the brief window $\gamma^*(t')$ suppresses the signal outside of a suitable neighborhood of t, the STFT is simply a "local spectrum" of x(t') around time t. The STFT is clearly a linear TFR and, generally speaking, is a complex function. Among his characteristics we can note that, fixed a signal x(t'), the corresponding $STFT_X^\gamma (t, f)$ is a function of the specific analysis window $\gamma^*(t')$, moreover STFT maintains also frequencial and temporal shifts in the signal x(t'):

$$\tilde{x}(t) = x(t) e^{j2\pi f_0 t} \Rightarrow STFT_{\tilde{X}}^\gamma (t, f) = STFT_X^\gamma (t, f - f_0),$$

$$\tilde{x}(t) = x(t - t_0) \Rightarrow STFT_{\tilde{X}}^\gamma (t, f) = STFT_X^\gamma (t - t_0, f) e^{-j2\pi f t_0} \quad (1.7).$$

The STFT can be expressed also by mean of the signal spectra and window spectra:

$$STFT_X^{(\gamma)} (t, f) = e^{-j2ft\pi} \int_{f'} X(f') \Gamma^*(f'-f) e^{j2f't\pi} df' \qquad (1.8).$$

Ignoring the phase factor $e^{-j2ft\pi}$, this expression in the frequency domain is analogous to the expression (1.6) in the time domain, in fact it shows that the STFT can also be interpreted as the IFT of the "spectrum window" $X(f') \Gamma(f'-f)$, in which the window spectrum $\Gamma(f)$ is simply the FT of the temporal window $\gamma(t)$.

Since the STFT evaluated at t time is the spectrum of the signal x(t') multiplied by the window $\gamma^*(t'-t)$, it shows all the characteristics of the signal contained within the range of duration of the window in a neighborhood of t time. It is clear that to obtain a good time resolution SFTF it requires a short window $\gamma^*(t')$, while a good frequency resolution requires a narrow-band analysis window, so it requires a quite long window. Unfortunately, the Uncertainty Principle denies the existence of windows with arbitrarily small bandwidth and duration, so the time-frequency resolution of STFT is necessarily limited. Therefore, there are two conflicting demands for the resolution of STFT: improving temporal resolution, using a short window, involves the deterioration of the frequency resolution and vice versa. Finally, let us consider two extreme choices of the analysis window. The first is the perfect time resolution which requires choosing as a window the Dirac pulse that is infinitely narrow:

$$\gamma(t) = \delta(t) \Rightarrow STFT_x^{(\gamma)}(t, f) = x(t)e^{-j2ft\pi} \qquad (1.9).$$

In this case, the STFT reduces to the signal x(t) and, while retaining all the changes in temporal resolution, does not display any frequency resolution. The second provides a perfect frequency resolution by a constant window:

$$\Gamma(f) = \delta(f) \Rightarrow STFT_x^{(\gamma)}(t, f) = X(f) \qquad (1.10).$$

The STFT of x(t), in this case is reduced to its FT and therefore has no temporal resolution.

1.4.2 THE WAVELET TRANSFORM

Another important linear TFR is the time-frequency version of the *wavelet transform* (WT) or *ondina transform* defined as follows:

$$WT_X^{(\gamma)}(t, f) = \int_{t'} x(t') \sqrt{\left|\frac{f}{f_0}\right|} \gamma^* \left(\frac{f}{f_0}(t'-t)\right) dt' \qquad (1.11),$$

where $\gamma(t)$, called wavelet of analysis, is a band-pass function real or complex, centered at t=0 in the time domain. The parameter f_0 is the center band frequency of the $\gamma(t)$.

The WT was originally introduced as a representation of a time scaled signal and this formulation can be regained by the time-frequencial expression (1.11) considering the scaling parameter as $\alpha = \frac{f_0}{f}$.

WT retains both the temporal translation and the temporal scaling:

$$\tilde{x}(t) = x(t - t_0) \Rightarrow WT_{\tilde{X}}^{(\gamma)}(t, f) = WT_X^{(\gamma)}(t - t_0, f) \qquad (1.12),$$

$$\tilde{x}(t) = \sqrt{|a|} x(a t) \Rightarrow WT_{\tilde{X}}^{(\gamma)}(t, f) = WT_X^{(\gamma)}(a t, \frac{f}{a}) \qquad (1.13),$$

while it does not preserve the frequencial shift.

Similarly, to the STFT, the WT can be interpreted, for each frequency f of analysis, as the output of a band-pass filter centered in f. But whereas in the case of STFT the bandwidth of the band-pass filter is independent from the central analysis frequency f, in the case of WT the bandwidth of the filter is proportional to f or, equivalently, the figure of merit Q of the filter (Q = band center frequency/bandwidth) is independent from f.

In principle, the WT is affected by the same restrictions as the STFT in fact the time and frequency resolutions cannot be improved simultaneously. The difference is that

12

while the STFT has a time-frequency resolution independent from the frequency of analysis, the WT analyzes the higher frequencies with a better time resolution but a worse frequency resolution.

The existence of a synthesis formula for the linear TFR just seen, allows the reconstruction of a signal starting from its STFT or from WT. This makes them suitable for many applications such as analysis of time-varying signal, identification of systems, spectral estimation, the filtering time varying, the signal and image processing, the acoustic and seismic signal processing and so on.

1.5 THE QUADRATIC TIME-FREQUENCY REPRESENTATIONS

1.5.1 GENERALITIES

Although the linearity is a very desirable characteristic, very desirable is too the quadratic structure of a TFR if we want it to be interpretable as the time-frequency energy distribution or spectrum of instantaneous power, since energy is a quadratic representation of signal. An "Energetic" TFR combines the concepts of instantaneous power $p_X(t) = |x(t)|^2$ and of spectral density of energy $p_X(f) = |X(f)|^2$. This interpretation of energy is expressed by *marginal properties*:

$$\int_f T_x(t, f)df = p_X(t) = |x(t)|^2 \qquad (1.14)$$

$$\int_t T_x(t, f)dt = P_X(f) = |X(f)|^2 \qquad (1.15),$$

13

according to which the one-dimensional energetic density $p_x(t)$ and $P_x(f)$ are "marginal density" of the TFR $T_x(t,\ f)$. As a result of that the energy of signal x(t):

$$E_x \ = \ \int |x(t)|^2 dt \ = \ \int |X(f)|^2 df \qquad (1.16),$$

can be derived by integrating $T_x(t,\ f)$ along the entire time-frequency plane.

The marginal properties puts in connection the integral of time and frequency of the TFR with the respective energetic densities $|X(f)|^2$ and $|x(t)|^2$ but they do not provide an interpretation of $T_x(t,\ f)$ as a "time-frequency density of Energy" for each point of time-frequency plane. In fact, the uncertainty principle makes it impossible to talk about energy at a given time and a certain frequency.

Many TFR can be interpreted roughly in terms of signal energy, even if they do not satisfy the marginal properties. Two important examples are the *spectrogram* (SPEC) and the *scalogram* (SCAL), defined as the square of the modules of the linear TFR seen above:

$$SPEC_x^{(\gamma)}(t,\ f) \ = \ \left| STFT_x^{(\gamma)}(t,\ f) \right|^2 \qquad (1.17),$$

$$SCAL_x^{(\gamma)}(t,\ f) \ = \ \left| WT_x^{(\gamma)}(t,\ f) \right|^2 \qquad (1.18).$$

The spectrogram has been used extensively to analyze speech signals and other non-stationary signals while the scalogram can be viewed as a "constant Q" version of the spectrogram.

In addition to the "energetic" interpretation of the quadratic TFR there is another in terms of correlation functions. A "correlative" TFR $T_x(\tau,\ v)$ combines the temporal correlation $r_x(\tau)$ and spectral correlation $R_x(\nu)$, which are both quadratic signal

representations. This representation is expressed by "*correlative marginal properties*":

$$T_x(\tau,0) = r_x(\tau) = \int_t x(t + \tau)x^*(t)dt$$

$$T_x(0, \nu) = R_x(\nu) = \int_f X(f + \nu)X^*(f)df \qquad (1.19),$$

where the variables τ and ν are the temporal and frequency lag, respectively.

1.5.2 THE QUADRATIC SUPERPOSITION PRINCIPLE

The spectrogram of the sum of two signals $x(t)+y(t)$ is not simply the sum of the

two spectrograms $SPEC_x^{(\gamma)}(t, f) + SPEC_y^{(\gamma)}(t, f)$, in fact the linear structure of the

STFT is violated by the definition of SPEC. Each quadratic TFR satisfies the "*quadratic*

superposition principle":

$$x(t) = c_1 y(t) + c_2 z(t) \Rightarrow$$

$$T_x(t, f) = \left|c_1\right|^2 T_y(t, f) + \left|c_2\right|^2 T_z(t, f)$$

$$+ c_1 c_2^* T_{yz}(t, f) + c_2 c_1^* T_{zy}(t, f) \qquad (1.20),$$

where $T_x(t, f)$ is the "auto-TFR" of the signal $x(t)$ and $T_{yz}(t, f)$ is the "cross-TFR" of

the two signals $y(t)$ and $z(t)$. If we generalize the quadratic superposition principle to a

signal with N components $x(t) = \sum_{k=1}^{N} c_k x_k(t)$ we obtain the following rules:

* For each component signal $c_k x_k(t)$ there is a "term signal" or "auto-component"

$\left|c_k\right|^2 T_{x_k}(t, f)$;

* For each pair of component signals $c_k x_k(t)$ and $c_1 x_1(t)$, with $1 \neq k$, there is an "interference term" or "cross-term" $c_k c_1^* T_{x_k, x_1}(t, f) + c_1 c_k^* T_{x_1, x_k}(t, f)$.

Therefore, for a signal with N components x(t), the TFR will include N terms signal and $\binom{N}{2} = \frac{N(N-1)}{2}$ interference terms, a number that increases as the square of N.

The interference terms of the spectrogram and of the scalogram are oscillatory structures located in regions of time-frequency plane where the terms of signal overlap. Therefore is valid the important property that if two signal components are sufficiently far apart, on the time-frequency plane the corresponding interference terms will be zero. The drawback of the spectrogram and of the scalogram consists mainly in the lack of concentration or time-frequency resolution.

1.5.3 THE WIGNER DISTRIBUTION AND THE AMBIGUITY FUNCTION

Among the quadratic TFR with an energetic interpretation there is the Wigner distribution (WD) which exhibits an excellent time-frequency concentration and a series of favorable mathematical properties, although if shows some drawbacks like relevant terms of interference:

$$WD_{x, y}(t, f) = \int_\tau x(t + \tfrac{\tau}{2}) y^*(t - \tfrac{\tau}{2}) e^{-j2f\tau\pi} d\tau$$

$$= \int_v X(f + \tfrac{v}{2}) Y^*(f - \tfrac{v}{2}) e^{j2fv\pi} dv \qquad (1.21).$$

Among properties is included that the WD preserves the time shifts and the frequency shifts and satisfies the marginal properties (1.14) and (1.15), or, in other words, that the integration by frequency and by time of the WD of a signal correspond respectively to its instantaneous power and to its density spectral energy. Therefore the WD can be interpreted as two-dimensional energy distribution of a signal on the time-frequency plane, although as already stated, the uncertainty principle forbids being regarded as a time-frequency energy punctual density.

Among the correlative TFR an equally important role occupies the *ambiguity function* (AF):

$$A_{x,y}(\tau, v) = \int_t x(t + \tfrac{\tau}{2})y^*(t - \tfrac{\tau}{2})e^{-j2\pi vt}dt$$

$$= \int_f X(f + \tfrac{v}{2})Y^*(f - \tfrac{v}{2})e^{j2\pi\tau f}df \qquad (1.22).$$

The AF can be interpreted as a time-frequency correlation function. The WD and AF are dual in the sense that they are a pair of Fourier transforms:

$$A_{x,y}(\tau, v) = \iint_{tf} WD_{x,y}(t, f)e^{-j2\pi(vt-\tau f)}dtdf \qquad (1.23).$$

Often the characteristics of the interference terms (ITs) of the WD generate problems in the applications with multi-component signals, in fact, while the ITs of the spectrogram or scalogram are zero, if the corresponding terms of the signal do not overlap, the ITs of the WD are present regardless from distance among the signal components. From a practical point of view, the ITs are a problem because they can overlap to the auto-terms making it difficult the diagrams interpretation of the WD and of the AF

1.5.4 THE TIME-FREQUENCY REPRESENTATIONS INVARIANT TO TRANSLATION: THE COHEN'S CLASS.

In addition to WD, there are many other TFR with an energetic interpretation and many of them satisfy the property of invariance to time and frequency shifts or "covariance": if the signal x(t) is delayed in time or shifted in frequency then its TFR will be translated the same time delay or frequency modulated:

$$\tilde{x}(t) = x(t - t_0)e^{j2\pi f_0 t} \Rightarrow T_{\tilde{x}}(t, f) = T_x(t - t_0, f - f_0) \qquad (1.24).$$

The class including all the quadratic TFR which are invariant to time and frequency shifts is known as the *quadratic Cohen's class*. Major elements of this class are the spectrogram and the Wigner distribution.

Each member of Cohen's class can be interpreted as a WD bi-dimensionally filtered, in fact it can be shown that a TFR, $T_x(t, f)$ is a member of Cohen's class, denoted by C_E, if and only if it can be derived from the WD of the signal x(t) by a time-frequency convolution:

$$T_x \in C_E \Leftrightarrow$$

$$T_x(t, f) = \iint_{t'f'} \Psi_T(t - t', f - f')W_x(t', f')dt'\,df' \qquad (1.25)$$

Then each member C_E is associated with a single kernel function $\Psi_T(t, f)$, or bi-dimensional filter, independent from the signal. Of course the previous convolution (1.25) becomes a simple multiplication in the domain of the Fourier transform.

1.5.5 THE SPECTROGRAM AND THE SMOOTHED PSEUDO WIGNER DISTRIBUTION

Although the WD is theoretically interesting for its mathematical properties, its practical application is often limited by the presence of interference terms. The latter have an oscillatory nature therefore can be attenuated by a smoothing operation (low-pass filtering). According to (1.25) each TFR of Cohen's class can be derived from the WD by mean of a convolution with a kernel function $\Psi_T(t, f)$. Of course this convolution will produce a smoothing of the WD, or a bi-dimensional low-pass filtering, only if $\Psi_T(t, f)$ is a kernel function provided with sufficient smoothing properties. In this case we will define the resulting TFR as *smoothed Wigner distribution* (SWD) and the kernel $\Psi_T(t, f)$ as the *smoothing function*. Unfortunately, this attenuation of interference terms is payed with the loss of time-frequency concentration because the smoothing typically involves the spectral enlargement of the signal terms of the WD. Therefore, there is an inverse relationship between good interference attenuation and a good time-frequency concentration, the trade-off between time and frequency resolution. A smoothing function $\Psi_T(t, f)$ over a wide domain of WD, which correspond to a low-pass filter with narrow band, produces a good interference attenuation but worsens the time-frequency concentration and vice versa.

Among the SWD showing shifts invariance deserves particular attention the classical spectrogram that can be expressed also as:

$$SPEC_X^{(\gamma)}(t, f) = \iint_{t'f'} WD_\gamma(t'-t, f'-f)WD_X(t', f')dt'\, df' \qquad (1.26).$$

Note that the spectrogram is nothing else that a SWD with a smoothing function

$\Psi_{SPEC}(t,\ f)\ =\ WD_\gamma\,(-t,-f)$ or, except for the reversal of the axis, the WD of the analysis window $\gamma(t)$ of the spectrogram. This means that the full extent $\Psi_{SPEC}(t,\ f)$ cannot be less than the minimum prescribed by the uncertainty principle. Consequently, the spectrogram smoothing is relevant, like relevant is also the attenuation of the interference terms, but this implies a lack of concentration as previously seen.

This inverse relationship between interference attenuation and concentration (time-frequency resolution trade-off) is overcome by the *smoothed pseudo Wigner distribution* (SPWD):

$$SPWD_X^{(g,\ H)}(t,\ f)\ =\ \iint_{t,f} g(t-t')H(f-f')WD_X(t',\ f')dt'\ df' \qquad (1.27),$$

which allow us to adjust freely and independently the smoothing effect Dt and Df. The SPWD is defined by a function $\Psi_{SPWD}(t,\ f)\ =\ g(t)H(f)$ that can be separated where g(t) and H(f) are two windows whose lengths independently determinate the extent of smoothing Df and Dt. This brings advantages in practical applications where the decoupling of temporal and frequency smoothing results in greater flexibility of use and computational efficiency. The special case of SPWD corresponding to the choice where g(t)=d(t), or in other words to the choice where Dt=0, is defined *pseudo Wigner distribution* (PWD). The PWD is essentially a "short-Wigner distribution" which uses a mobile analysis window.

Worth a short mention the fact that new smoothed Wigner distributions are currently under development. Examples of SWD recently defined are the *cone kernel distribution*, the *generalized exponential distribution*, the *Butterworth distribution*, the *Choi-Williams distribution* [1]. These are all SWD and show a smoothing effect more complicated than that used for SPWD; in fact these smoothing act in order to mitigate the interference terms and simultaneously preserving the terms of the signal and the mathematical favorable properties.

From this brief discussion it shows that the time-frequency representations are powerful tools for analysis and processing of non-stationary signals for which a separate analysis in the time domain and frequency domain would be inadequate. Applications of the TFR are found in the fields of quantum mechanics, optics, acoustics, radar, sonar, radio astronomy, communications, bioengineering, etc.. There is not an optimum TFR for any signal-processing problem but the choice of the specific TFR and the choice of his smoothing depend upon the specific application. We will take care particularly about applications relating to biological ultrasound signals for which, so far, the best results have been obtained adopting the spectrogram and the pseudo Wigner distribution.

CHAPTER 2

SOME PROCESSING TECHNIQUES OF ULTRASOUND DOPPLER SIGNALS

2.1 GENERALITIES

Modern medicine uses a lot of ultrasonic investigations, both in clinical than in experimental applications, because this technique has the important characteristics of being non-invasive, non-traumatic, and, at levels of intensity commonly used, not harmful to biological tissues. The ultrasonic investigation involves a lower risk for tissues injury than the traditional X-ray diagnostic technique, which remains anyway a complementary clinical method, and for this reason the ultrasonic investigation is often used to infants diagnosing, children and pregnant women. Moreover, with some ultrasound techniques, we can obtain accurate information with spatial resolution much higher than those obtained with the radiographic method, or with computerized axial tomography (TAC) or nuclear magnetic resonance (NMR).

The most common technique employed is the "eco-pulse" one, which produces ultrasound images by mean of the detection of reflected echoes, through changes in acoustic impedance on the interfaces encountered by a train of ultrasonic waves on its way through the tissue under examination.

Traditionally, the importance of ultrasound images derives from the morphological information that they contain. Subsequently the ultrasound have been used to obtain qualitative information regarding quantitative and even structural information about biological tissues. For example, you can differentiate benign and malignant tissue according to their acoustic properties and to the internal structural organization, or you can calculate volumes or take measures.

The applications of ultrasound are also found in medical therapy [7]. The clinical chemotherapy is often limited by the poor selectivity that cytotoxic drugs exhibit against tumor tissues and normal tissues. When it is possible to identify a well-defined tumor region, for local therapeutic treatments are used methods such as ionizing radiation, hyperthermia, etc.. Recent laboratory studies have revealed the action of enforcing and the ability to activate on some drugs by unfocused ultrasound both continuous and pulsed.

The ultrasound information can also be used to perform functional assessments. For example, you can use the M-mode paths of the motion of the heart valves in order to obtain data on valves disease and about cardiac functions. You can also analyze the frequency modulation of the Doppler signals coming from blood flow in order to classify a disease or circulatory functions or in order to check the heart valves.

The Doppler technique is used in medicine in order to measure for blood flow velocity V using the following equation:

$$V = \frac{Cf_D}{2f_0 \cos \theta} \ (2.1),$$

where f_0 is the wave frequency emitted by the transducer, C is the ultrasonic wave propagation velocity within the medium under analysis, f_D is the frequency shifts for the Doppler effect, θ is the angle between the incident ultrasound beam and the motion direction of the flow.

Equipment based upon the use of Doppler techniques allows obtaining information about the movement of blood flow within the vessels, in the form of tracks, images and auditory signals. In relation to the different methods of acquisition and

23

presentation of information, to the diagnostic applications and to the complexity, we can differentiate the devices into four classes. For the Doppler equipment based on *continuous wave* (CW) the structures, which are scanned by the acoustic beam simultaneously, contribute to the formation of the Doppler signal. In those based on the *pulsed wave* (PW), the transmission of short pulses allows the localization of the depth to which the measure of the speed refers to, allowing the acquisition of local speeds, allowing the definition of the position of blood vessel walls and allowing the analysis of the velocity profile along a vascular diameter, through a synchronous acquisition at different depths. This requires that the sample volume to be moved along the line of inquiry, such as in *mono-gate* systems where it is not possible to obtain real-time measurements. Alternatively, the reconstruction of the velocity profile can be created automatically using *multi-gate* systems. The *Duplex Scanner* equipments belong to the third class and consist of an association of real time B-scan together with a transducer mono-probe, suitable for the acquisition of the Doppler signal and independently positioned at an appropriate angle. The combined use of both techniques allows a substantial increase in diagnostic power and a considerable reduction of the time required for performing the examination. The fourth class of equipments includes the *color Doppler* that, in principle, is similar to the Duplex Scanner. In the Color Doppler the analysis is repeated over N points each line for M rows. The result is an array made of N*M points which corresponds to a Doppler image of the speed, coded by mean of colors, which overlaps with the traditional morphological black and white image. With this technique anatomical information are associated with the functional information extended to the entire image or to only a region of interest (ROI).

In the initial phase of the of ultrasound diagnostic applications, many researchers have investigated and have used methods used in the time domain in order to extract the frequency or, equivalently, the information of blood velocity from Doppler signals transmitted in the irradiated tissues. Following the appearance of real-time spectrum analyzers, these methods have lost much of their importance. The advent, then, of the Color-Doppler techniques has given new urgency to the time domain analysis as computationally less expensive and therefore sometimes necessary to obtain images in real time. Over the past decade then have been established new techniques for the analysis and representation of pulsed Doppler signals among which there is one that uses the time-frequency representations (TFR).

2.2 TIME DOMAIN METHODS

The methods of analysis of the Doppler signal in time domain, for example the ones that use a function of time to represent the signal, have been the first to be adopted, since the early '60s, to estimate the average frequency, its mean square value or the maximum frequency shift of the Doppler signals.

An early example of pulsed Doppler blood flow detector using pulsed ultrasound dates back to 1970 and is the work of DW Baker [1].

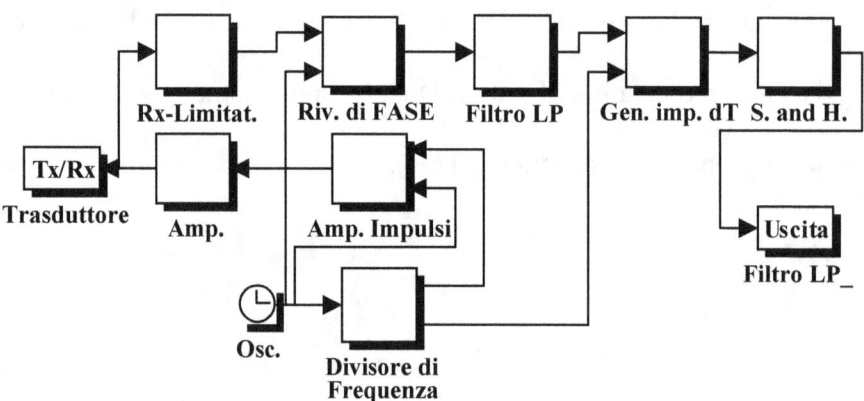

Fig 1. Pulsed Doppler blood flow detector.

The simplified block diagram of fig. 1 shows the operating principle of an early type of coherent Doppler detectors. The echoes produced by the discontinuity of moving surfaces are modulated both by phase and by amplitude. A system that allows detecting the presence of Doppler effect by comparing the waveform coming from a sample volume with a replica of the transmitted signal through a phase comparator is called *coherent*. In this system there is a pilot oscillator (Osc.) that produces the pilot carrier wave for the transducer excitation, for the coherent demodulation of the received echo and for the timing of the various blocks. This operation is performed by the frequency divider which generates the frequency repetition of the transmitted pulse and generates the sampling frequency of the received and demodulated signal. We find, then, the generator of ultrasonic pulses, controlled by the frequency divider, and the linear amplifier to adjust the output power coming out from the transducer. Within the time intervals in which the transducer works as for reception, the signal is picked up by a high frequency amplifier which is able to amplify it to a level that can be send as an input to a phase detector which compares the instant phase difference between received echoes and transmitted echoes. The signal produced by the phase detector, which contains the Doppler information concerning the sample volumes where the echoes go through during the interval of listening, is acquired by a sample & hold (S&H) with a sampling frequency, which determines the number, and the depth of sample volumes examined. Finally the Doppler signal is amplified in order to be interpreted by acoustic analysis or by graphical analysis. We note that this system, for the poor accuracy characteristics it shows, appears to be a detector but not a blood flow meter, besides it does not provide the ability to determine the direction of flow. The maximum detectable Doppler shift is equal to the half of the pulse repetition frequency, which is in other words the sampling rate for the well-known Nyquist limit. The operations mode are two, one in which we investigated the individual sample volume limiting the range of integration of S&H to the minimum duration, the other one in which this interval is extended to all the sample volumes that belong to a blood vessel in order to deduce an average speed information. The version of such a system that provides multi-channel parallel detection allows both the measurement of vascular diameter and allows the representation of real time velocity profiles.

Another interesting system is the Doppler Topoflow presented by M. Brandestini [2] in 1978. This system, shown schematically in Fig. 2, is an instrument for directional blood velocity measurements using ultrasonic pulsed multi-gate type with a serial architecture.

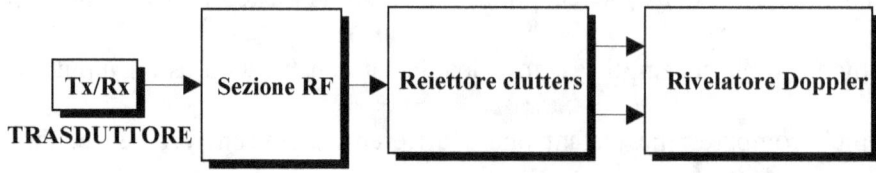

Figure 2. Block diagram of Doppler flow meter Topoflow.

The method of signal serial processing is able to implement, effectively, a multi-gate Doppler system in which sample volumes, considered simultaneously, are in large number. Under this concept, the blocks of signal processing are shared sequentially by signals coming from different depths. This method has been developed since the mid-'70s, inspired by the concept of moving target indicator (MTI) coming from the field of radar applications.

The Fig. 3 shows the block diagram of a radio frequency section of the system which is based on a scheme of coherent demodulator. Short ultrasonic pulses, generated by the choking of the central oscillator, are amplified and transmitted into the tissue by the piezoelectric transducer at the pulse repetition frequency (PRF) which determines the performance of the system.

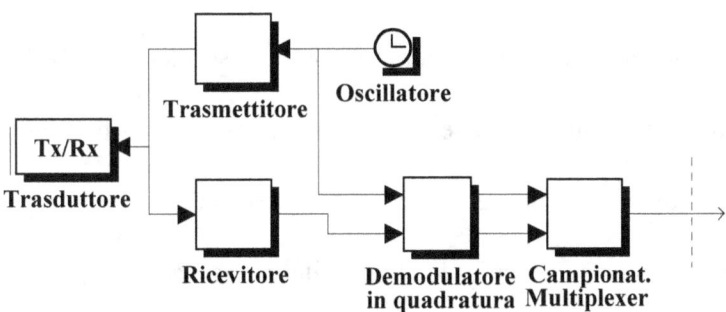

Fig 3. Block diagram of a radio frequency stage.

During the listening intervals, the crystal acts as a receiver and picks up the signal spread from the interfaces and from the mobile particles. The echoes are amplified and mixed with the signal produced by the central oscillator in order to detect the phase changes occurred. This is achieved through a coherent demodulator in quadrature phase, whose diagram is shown in fig. 4, which allows discriminating the direction in which the flow moves from the transducer.

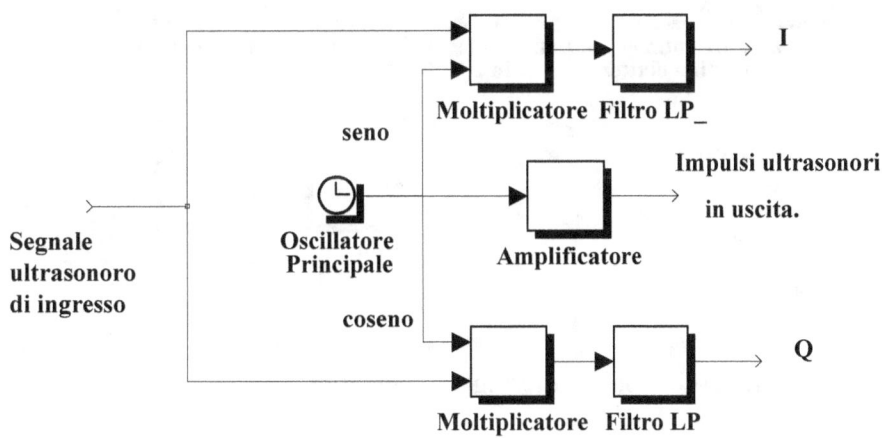

Fig 4. Quadrature-phase demodulator Block diagram.

The output signal from the quadrature demodulator contains information about the Doppler shifts but also other unwanted components, constituted mainly by the echoes produced by various discontinuity surfaces such as vessel walls. In situations like those of clinical use, the amplitude of the noise components exceeds that of the Doppler signal of interest of even two orders of magnitude. Fortunately, the spectral components of the Doppler signal and the noise are separated; in fact, the latter is concentrated at low frequencies because they correspond to obstacles that are moving slowly or stationary. In the case of pulsating flow, the movement of blood vessel walls produces frequency components which are considered stationary below 100 Hz It becomes therefore necessary to filter such noise components occurring in numerical form, as is shown in the fig. 5, after converting the signal in that form. At this point, the two-quadrature channels, numerically filtered serially, are separated again for the revelation of speed.

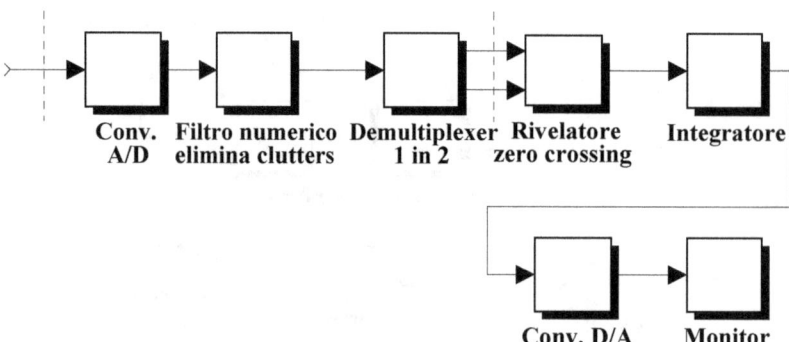

Fig 5. Section of Doppler filtering and detection.

The speed estimation was implemented using a zero crossings counter (zero-crossing). This system produces, for each predetermined interval estimation, an output which is proportional to the number of zero crossings of the input signal.

This system, though interesting for the implementation of the serial multi-gate technique, has major errors in the frequency estimating, due to the sensitivity of the zero-crossing technique to the noise and to the aliasing phenomena for the Doppler spectrum, which are present well below the Nyquist rate.

Of great interest, both theoretical and applied, appears to be the autocorrelation technique. The work of C. Kasai dated 1985 [4], presents a system capable of producing two-dimensional images of blood flow in real time, using the autocorrelation technique applied to pulsed ultrasound. The operating principle is based on a radial scanning of the section of interest by an ultrasonic beam; each radial vector, obtained in such a way, is examined by a small number of ultrasonic pulses so that the estimation of the speed parameters, using the auto-correlation technique, can be obtained at various depths. These parameters, then, are coded by colors and the resulting image is overlapped to the one in gray scales, representing the morphological information. Such a system is called color Doppler.

The Fig.6 shows the block diagram of the system of two-dimensional visualization of blood flow using the autocorrelation technique. The system is equipped with a conventional B-mode unit of image and with a unit for spectral analysis of each sample volumes, by mean of a fast Fourier transform (FFT). The oscillator (Osc) is a high-frequency oscillator whose output is divided appropriately (Div) in order to produce the continuous wave needed to demodulate the Doppler signals and in order to enable synchronization of the various blocks.

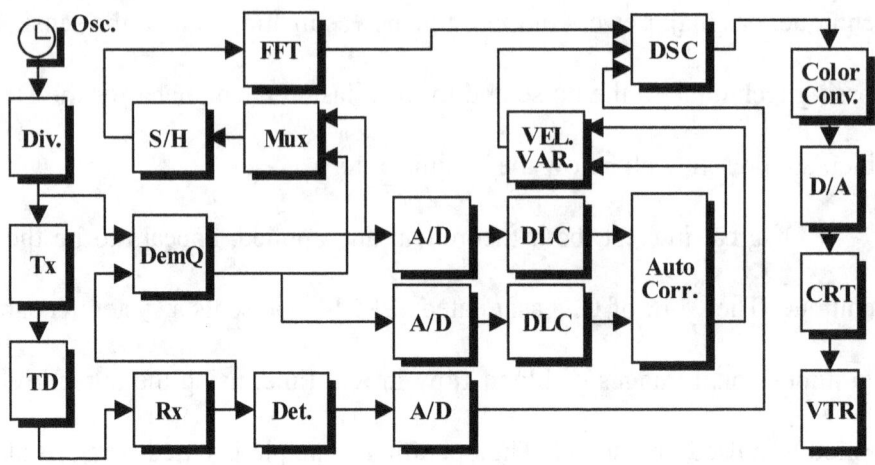

Figure 6. Block diagram of the system Doppler autocorrelation.

The signals received by the transducer (TD) are first amplified (Rx) and then are sent to a synchronous demodulator in quadrature-phase (DemQ), where the phases of the reference frequencies differ by 90°. The demodulation is a coherent type, then the output signal from the two low-pass filters in the quadrature demodulator contains the complex Doppler frequencies which are produced by the Doppler effect. The pair of signals in quadrature, or in other words, the complex envelope of the signal received by the transducer, after the digital form conversion (A/D), are sent to the filters (Delay Line Canceler) in order to eliminate the echoes coming from stationary tissues or from the slowly moving tissues, and from the DLC to the complex autocorrelator described in Fig.7.

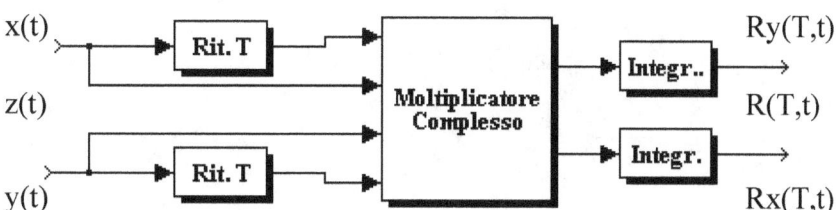

Figure 7. Complex Autocorrelator for the R(T) calculation.

The estimation of the average Doppler frequency and of its variance, which respectively

indicate the speed and the turbulence of blood flow within the sample volume, can be

obtained from the autocorrelation function R(t) of the complex envelope z(t)=x(t)+jy(t) of

the ultrasound signal E(t) which is received by the transducer [10]. It is possible to show

that the estimation of the mean frequency and of the variance can be obtained, in an

approximate but very efficient way from the computational point of view, from the

module |R(T)| and from the phase Φ(T) of the autocorrelation evaluated at t=0 and t=T,

where T is the interval emission of ultrasonic pulses, as shown in Appendix A.

$$\varpi = \dot{\phi}(0) \cong \frac{\phi(T)}{T} \qquad (2.2),$$

$$\sigma^2 \cong \frac{2}{T^2}\left\{1 - \frac{|R(T)|}{R(0)}\right\} \qquad (2.3).$$

The Fig. 7 is just a scheme for real time calculation of R(T) and of Φ(T). The real and

complex components of the complex Doppler signal z(t) are sent to a complex conjugate

multiplier together with their delayed versions of T. The autocorrelation is obtained, then, by integrating the product for a certain period from which depends the accuracy of the estimation.

$$R(T, t) = \int_{t-nT}^{t} z(t') \cdot z^*(t'-T)dt'$$

$$= R_x(T, t) + jR_y(T, t) \qquad (2.4).$$

In the formula (2.4) n represents the number of ultrasonic pulses which are transmitted consecutively in the same direction, then nT represents the integration time for each vector of the scanning. Once obtained R(T,t) it follows easily the calculation of the remaining terms:

$$|R(T, t)| = \sqrt{R^2_x(T, t) + R^2_y(T, t)} \qquad (2.5),$$

$$\phi(T, t) = arc\,tan\,\frac{R_y(T,t)}{R_x(T,t)} \qquad (2.6),$$

$$R(0, t) = \int_{t-nT}^{t} z(t') \cdot z^*(t')dt' \qquad (2.7).$$

From these expressions it is clear that the output of the autocorrelator is a function of both time, or in other words is function of the particular sample volume chosen, and is function of the duration of the integration time. The latter cannot be stretched beyond measure in order to improve the estimation, because this would reduce the number of images per second and therefore the ability of the system to display in real-time.

Coming back to the diagram shown in Fig. 6 it appears clear that the output of the autocorrelator give us the possibility to estimate, for the N sample volumes of the M-rays of scanning, the speed and the turbulence of the blood (Speed/Var.) which are captured by a digital scanner (DSC) and encoded using different colors. To these information are added those in B-mode or in M-mode, obtained by the traditional envelope detection methods (Det.) and by spectral analysis (FFT), then are stacked in a single image and successively converted into analog form (CRT) so that can be available on a color monitor (VTR). This technique, often referred into literature as a technique of estimation of the phase angle of the autocorrelation, represents a good compromise between the quality of the estimation of the average frequency and the estimation of the real time performance, allowing us to analyze each frequency in the interval of Nyquist [-PRF/2,+PRF/2], where PRF is the pulse repetition frequency [11]. The disadvantage of this technique is to show a high variance of the estimation which increases with increasing of the mean frequency and of the Doppler bandwidth of the signal. This is a phenomenon very negative when are analyzed signals with a low signal to noise ratio (SNR) as are those coming from small and deep blood vessels and the low-speed signals which are degraded from the rejection characteristics of the anti-clutter filters. Although the limitations of this dissertation do not allow us to extend beyond the discussion on the methods in the time domain, we can just say that recently have been obtained better results than the methods just viewed, using the adaptive estimators of the average frequency which adapt themselves to the specific characteristics of Doppler signals [11].

As conclusion of this exposition, which is not at all exhaustive, of some estimation techniques of the average frequency of the pulsed Doppler ultrasound signals, we can confirm the renewed importance that these techniques are assuming in the implementation of systems that work with the limitation of the real-time calculation.

2.3 THE FREQUENCY DOMAIN METHODS

The analysis of Doppler signals in the frequency domain has become feasible thanks to the techniques of digital signal conversion and to the low cost implementation of the Fast Fourier digital techniques. In recent years, moreover, many of the complex analog circuits for signal processing have been validly superseded by the use of digital signal processors (DSPs) for which the price/performance ratio is continuously improving. The systems implemented by these processors offer several advantages compared to the corresponding analog ones; for example, are easier to calibrate, less susceptible to aging and environmental variations, less sensitive to noise, and show a better flexibility toward applications. Ultimately, digital signal processors have enabled new applications or improvements both of the techniques of Doppler signal processing in the time domain than in the frequency domain.

The autocorrelation Doppler system, shown in the previous paragraph in Fig.6, presented by Kasai and others. [4], associates this technique in the time domain with the spectral analysis by mean of the fast Fourier transform. In fig. 8 we recall the explanatory scheme of the FFT block of the system which represents a good example of technique in the frequency domain.

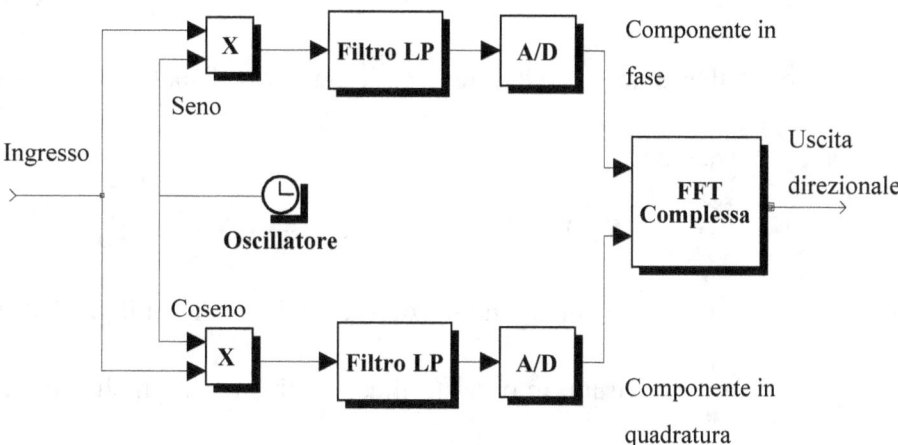

Figure 8. Block diagram of the complex FFT (CFFT) applied to the demodulated

signals in quadrature phase.

The use of the *complex Fast Fourier* (CFFT) allows to directly extracting the directional

information contained in the Doppler signal and it does not require further processing.

The complex Fourier transform (CFT) has some useful properties that allow it to detect

the direction of blood flow when applied to a complex Doppler signal whose real and

imaginary parts are in quadrature phase, i.e. if their phase difference is 90°. Consider a

complex signal, if its real part is even and its imaginary part is odd, then the CFT is real.

If, however, its real part is odd and its imaginary part is even, then the CFT is imaginary.

This property may be clarified by considering the example of a complex signal that

includes both cases:

$$x(t) = x_a(t) + x_b(t) = x_r(t) + jx_i(t)$$

$$x_a(t) = \left(A \cos 2\pi f_a t + jA \sin 2\pi f_a t \right)$$

$$x_b(t) = \left(B \cos 2\pi f_b t + jB \sin 2\pi f_b t \right),$$

where x_a and x_b represent, respectively, the first and second case. The Fourier transform of this signal is complex, omitting the calculations that can be found in Appendix A, shall be:

$$X(f) = A\delta(f - f_a) + jB\delta(f - f_b),$$

result that confirms the directionality of the CFT and the bilateral distribution of the information in its spectrum. In order to illustrate this result graphically, let us simulate a complex signal, which represents the two previous cases, and let us verify the validity of the properties of the CFT. In this regard we place:

$$x(t) = I + jQ = \{1000\cos(2\pi 1000 \cdot t) + 500 sin(2\pi 2000 \cdot t)\}$$
$$+ j\{1000 sin(2\pi 1000 \cdot t) + 500\cos(2\pi 2000 \cdot t)\},$$

which is the test signal whose phase and quadrature components are shown in fig.9 and in Fig.10 respectively. The signal, lasting 16 [ms], was simulated using a sampling frequency of 16 [kHz], then it has been represented by 256 samples. The spectrum of CFFT is shown in Fig. 11.

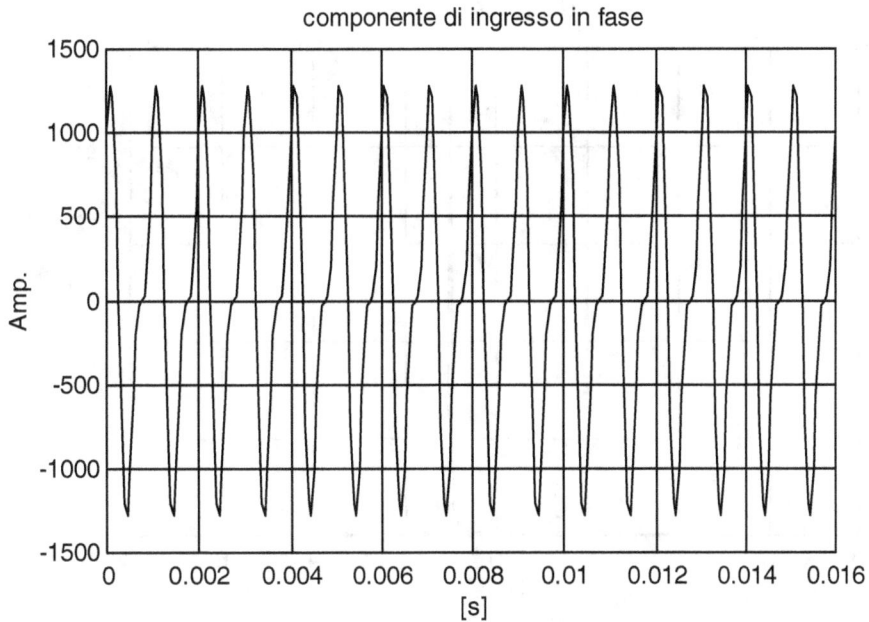

Figure 9. Diagram of the component in phase of test signal x(t).

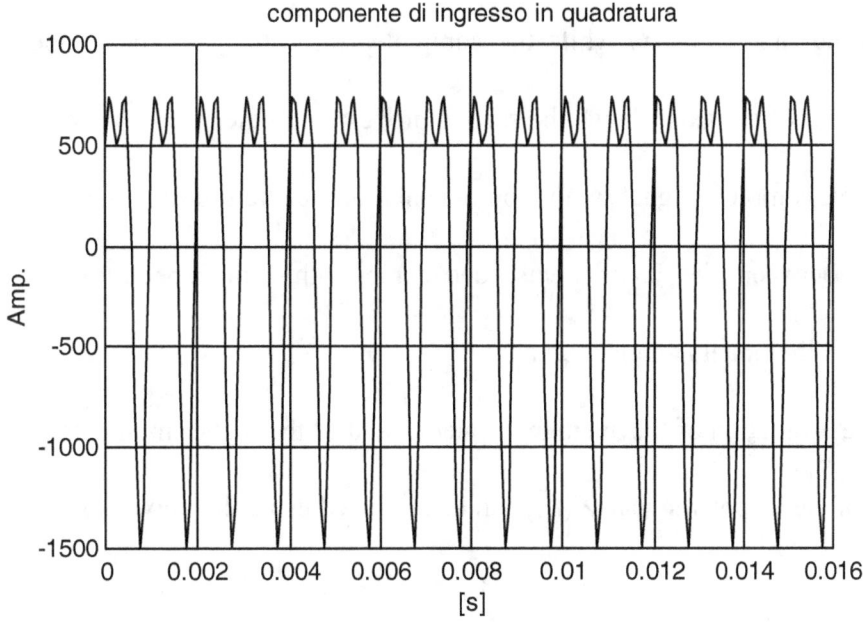

Fig 10. Diagram of the quadrature component of the test signal x(t).

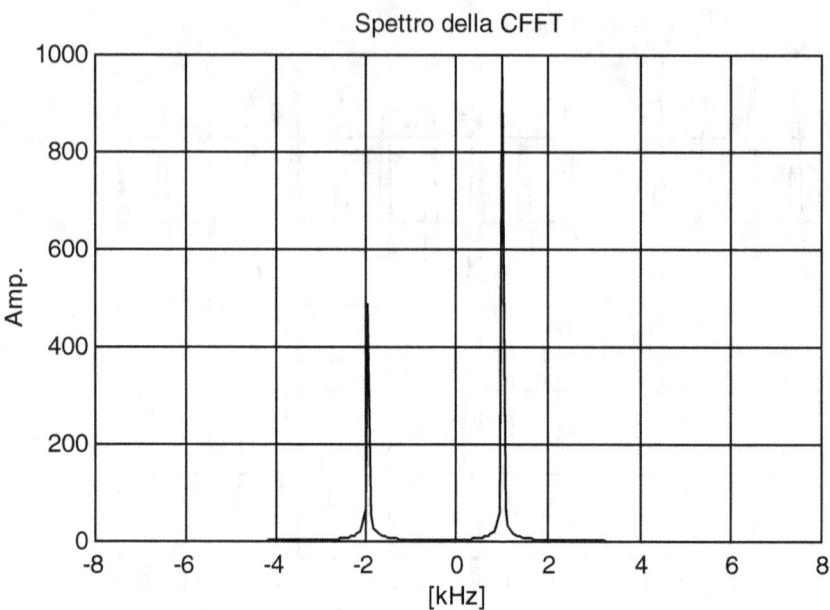

Fig 11. Spectrum of CFFT of the test signal x(t)

The frequency component at 1000 [Hz], located in the middle of the real spectrum,

illustrates the first property, while the component at 2000 [Hz], being located in the

imaginary half of the axis, shows the second property. In other words we can say that, in

the case of complex signal with components in quadrature, the CFFT makes the

separation between the components, detectable within the spectrum, allowing the

detection of directional Doppler signal.

The estimation of the average frequency and of the higher moments, knowing the

spectrum of the Doppler signal $X(f)$, can be produced using the power spectrum:

$$S(f) = X(f) \cdot X^*(f) \qquad (2.8),$$

as an approximation of the probability density function of frequency, namely:

$$p(f) = \frac{S(f)}{\int_{-\infty}^{+\infty} S(f)df} \qquad (2.9).$$

The physical interpretation of the (2.9) is that p(f)df represents the probability that the function x(t) could have an average power S(f)df in the frequency range (f,f+df). The denominator gives an appropriate normalization. On that basis, we can express the average frequency and the variance as:

$$\varpi \ = \ \int_{-\infty}^{+\infty} \omega p(\omega) d\omega \tag{2.10}$$

$$\mathrm{var}[\omega] \ = \ \int_{-\infty}^{+\infty} \omega^2 p(\omega) d\omega \ - \ \varpi^2 \tag{2.11}.$$

We stress finally that the simple estimators of the average frequency and of variance of the Doppler signal (2.10) and (2.11), compared with the estimators obtained in the time domain [3], exhibit a worst behavior in the presence of low levels of signal to noise ratio and result to be of more complex implementation and to be more expensive.

2.4 TIME-FREQUENCY REPRESENTATION METHODS

Over the past decade have established new techniques for the analysis and representation of pulsed Doppler signals among which there is one that uses time-frequency representations (TFR). The short Fourier transform (STFT) and its corresponding quadratic, the spectrogram (SPEC), are time-frequency representations and they provide the widely used tools for analysis and representation of the ultrasonic signals. The time-frequency resolution of such TFR, however, presents a significant limitation due to the uncertainty principle, which prohibits the unlimited increase of frequency and temporal resolutions, being their product bounded below. Several authors have also shown the limitations of the spectrogram as a tool for analysis of blood velocity

41

profiles [12]. In fact, the spectrogram, commonly used in Duplex Scanner systems is not able to detect the decelerations of the phase oscillations. Since the frequency of such decelerations is an important indicator of abnormal or turbulent flows generated by the presence of vascular stenosis, there is a need to develop different TFR which can be able to interpretate these variations and that could show better resolution characteristics.

Among the smoothed Wigner distributions (SWD), of which the spectrogram is a particular case, stands out because of its mathematical properties and characteristics of frequency concentration, the pseudo Wigner distribution (PWD). The smoothing function of the PWD, in fact, acts only in respect of the size of the frequency.

In later chapters, therefore, we will implement some efficient algorithms in order to calculate the spectrogram and the pseudo Wigner distribution and we will perform a performance comparison of their use as tools for analysis of Doppler ultrasound signals.

CHAPTER 3

AN EFFICIENT IMPLEMENTATION OF THE PSEUDO WIGNER-VILLE DISTRIBUTION

3.1 INTRODUCTION

The processing of a non-stationary signals by mean of the Fourier transform (FT) it produces very limited results so, in such cases, can be successfully used a TFR, for example, like the short Fourier transform (STFT) (1.6). An approximation of the spectral content, in the central point of the analysis window, can be obtained by mean of the corresponding quadratic TFR, the spectrogram (SPEC) (1.17), assuming that the signal x(t) could be stationary for the duration of the analysis window T. The time-frequency resolution of the STFT depends on the duration of the window T, in fact increasing the frequency it improves resolution but it decreases the ability of frequency tracking, or in other words increases the time interval between the estimated frequency and the next. The need for overcome of these limitations has led to investigate alternative representations of the signal and this has led to the identification of the pseudo Wigner-Ville distribution (PWVD) as a more powerful tool than the STFT in several applications. Many applications like the radar ones, sonar, seismic, biomedical, etc.. need for a real time calculation of the PWVD with widely varying resolution parameters, so we need to identify an optimized algorithm which allows an efficient and simple implementation of it.

3.2 THE PSEUDO WIGNER-VILLE DISTRIBUTION

If x(t) is a real signal, continuous and depending from the time variable t, the expression of its *Wigner distribution* (WD) (1.21), derived from the representation of the signal in time domain or in frequency domain, is:

$$WD_x(t, f) = \int_{-\infty}^{+\infty} x(t + \tfrac{\tau}{2})x^*(t - \tfrac{\tau}{2})e^{-j2\pi f\tau}d\tau \qquad (3.1)$$

$$WD_x(t, f) = \int_{-\infty}^{+\infty} X(f + \tfrac{v}{2})X^*(f - \tfrac{v}{2})e^{j2\pi tv}dv \qquad (3.2).$$

The elimination of the aliasing, as we shall see later, suggest us to use, instead of the signal x(t), the associated analytical signal:

$$z(t) = x(t) + jH[x(t)] \qquad (3.3),$$

where H[x(t)] is the Hilbert transform of x(t), obtaining a particular version of the WD that is called the *Wigner-Ville distribution* (WVD):

$$WVD_z(t, f) = \int_{-\infty}^{+\infty} z(t + \tfrac{\tau}{2})z^*(t - \tfrac{\tau}{2})e^{-j2\pi f\tau}d\tau \qquad (3.4).$$

The (3.4) implies that the evaluation of the WVD at time t is a transaction that is not causal because it requires the knowledge of the signal even for the times after t, and this does not make it suitable for the real-time applications. We can be overcome this limitation by applying the WVD to a windowed version of the signal. If the WVD is required at time t' then the windowed signal is:

$$x_w(t, t') = x(t)w(t - t') \qquad (3.5),$$

where w(t) is the window function symmetric with respect to t' which satisfies the relation: w(t)=0 for $|t| > \tfrac{T_w}{2}$. The WVD of the windowed signal is called the *pseudo Wigner-Ville distribution* (PWVD) and is described, starting from (3.4) as:

$$PWVD_z(t, f) = \int_{-T_w}^{+T_w} z(t + \tfrac{\tau}{2})z^*(t - \tfrac{\tau}{2})w(\tfrac{\tau}{2})w^*(-\tfrac{\tau}{2})e^{-j2\pi f\tau}d\tau \quad (3.6).$$

Using the properties of invariance for time shifts (Appendix A) the PWVD, at any time t', can be evaluated by shifting the signal x(t) so that the instant t' coincides with the origin, and then we get:

$$PWVD_z(0, f) = \int_{-T_w}^{+T_w} z(\tfrac{\tau}{2})z^*(-\tfrac{\tau}{2})w(\tfrac{\tau}{2})w^*(-\tfrac{\tau}{2})e^{-j2\pi f\tau}d\tau \quad (3.7).$$

It may also be rewritten as:

$$PWVD_z(0, f) = \int_{-T_w}^{+T_w} z'(\tau)w'(\tau)e^{-j2\pi f\tau}d\tau \quad (3.8),$$

where:

$$z'(\tau) = z(\tfrac{\tau}{2})z^*(-\tfrac{\tau}{2}) \quad\quad\quad\quad (3.9)$$

$$w'(\tau) = w(\tfrac{\tau}{2})w^*(-\tfrac{\tau}{2}) \quad\quad\quad\quad (3.10).$$

You can clearly see that (3.8) is similar to the definition of STFT (1.6), in fact it is equal to the FT of the signal z'(t) multiplied by the window w'(t). This implies that the resulting frequency resolution is determined by the spectrum of w'(t) and it is equal to that which is obtained using the same window for the calculation of the STFT. Since the multiplication in the time domain corresponds to the convolution in the frequency domain, the signals with symmetric spectrum, like the sine wave, will retain the spectrum shape if the FT of w'(t), i.e. W'(f), is an even function. For this to happen, for the symmetry properties of the FT, it is necessary that the actual window w'(t) has to be a real and symmetric function.

3.3 AN ALGORITHM FOR THE DISCRETE TIME PSEUDO WIGNER-VILE DISTRIBUTION.

In the practical applications the PWVD is calculated using his discrete-time version. The discrete equivalent of (3.6), which is called *discrete pseudo Wigner-Ville distribution* (DPWVD), is defined by:

$$PWVD_x(nT, f) = 2T \sum_{l=-L}^{L} x(nT + lT)x^*(nT - lT) \cdot$$

$$\cdot w(l)w^*(-l)e^{-j4\pi fl\tau} \qquad (3.11),$$

where T is the sampling period of time and w(lT)=0 when |l|>L, where L is a positive integer $(L \in Z^+)$. The properties of DPWVD are similar to those of continuous-time version, except for the periodicity in the frequencial variable. Exploiting the invariance for the shifts and standardizing the sampling period, T=1, then the (3.11) becomes:

$$PWVD_x(0, f) = 2 \sum_{l=-L}^{L} k(l) \cdot e^{-j4\pi fl}$$

$$\qquad (3.12),$$

$$k(l) = x(l)x^*(-l)w(l)w^*(-l)$$

where k(l) is called the kernel sequence of the DPWVD. As occurs:

$$PWVD_x(n, f) = PWVD_x(n, f + \tfrac{1}{2}) \qquad (3.13),$$

The DPWVD is periodic in the frequency domain with a period equal to the half the sampling period. If x(t) is a real signal with a bandwidth equal to B, in order to avoid the aliasing in the DPWVD then should happen that $f_s \geq 4B$. But if we consider the analytical signal z (n)=x(n)+jh[x(n)], where H[x(n)] is the Hilbert transform of x(n), although the frequency of DPWVD appear to be unchanged, the absence of the part of the spectrum corresponding to the negative frequencies allows to bring the limit for the elimination of the aliasing up to the frequency of Nyquist which is: $f_s \geq 2B$.

For the calculation of DPWVD using a digital computer we must also proceed to the discretization of the frequency. If we replace the variable f with a discrete variable $m\Delta f$, where $\Delta f = \frac{1}{N} = \frac{1}{2L+2}$, the (3.12) becomes:

$$PWVD_x(0, m\Delta f) = 2 \sum_{l=-L}^{L} k(l)W_4^{ml}$$

(3.14).

$$W_4 = \exp\left[\frac{-j4\pi}{N}\right]$$

To take advantage of an existing algorithm for the calculation of the FFT it is required that the index of the temporal sequence varies from 0 to N-1. This can be achieved by a periodic extension of the kernel sequence [20]:

$$k'(l) = \begin{cases} k(l), 0 \leq l \leq \frac{N}{2} - 1 \\ k(l - N), \frac{N}{2} + 1 \leq l \leq N - 1 \\ 0, l = \frac{N}{2} \end{cases}$$

(3.15),

where N=2L+2.

Now can rewrite the DPWVD as follows:

$$PWVD_x(0, m\Delta f) = 2 \sum_{l=0}^{N-1} k'(l)W_4^{ml}$$

(3.16),

that is a form that allows us to use an efficient FFT algorithm, whereas being $W_4 = \left(W_2\right)^2 = \left(\exp\left[\frac{-j2\pi}{N}\right]\right)^2$, the power of 2 represents a scale factor of the frequency axis by a factor of 2, and then in the calculation, it can be omitted.

3.4 OPTIMIZATION OF THE ALGORITHM FOR THE DISCRETE PSEUDO WIGNER-VILLE DISTRIBUTION CALCULATION

As seen in the previous paragraph the assessment of the DPWVD of a real signal x(t) can be done by a sequence of processing in which has been deliberately added a FFT algorithm, according to the structured programming methodology. This methodology requires the design of algorithms with independent and interchangeable functional modules, and it is well suited for the implementation that uses a digital signal processor (DSP). We can find a first version of the algorithm for the calculating of the DPWVD the following sequence of operations:

* Evaluation of the analytical signal z(t) from the signal x(t) under analysis

* Segmentation of the analytical signal z(t) according to the desired time resolution

* Calculation of the kernel sequence k'(l) for each segment

* Calculation of the FFT of the sequences.

In Appendix B it is shown the program **PWVD.m** that performs the algorithm implemented in MATLAB [21].

The symmetry properties of the Wigner kernel sequence, also valid for complex sequences such as the analytic signal, allow us to perform a significant optimization of the algorithm for the calculating of DPWVD. We have already shown that the actual window $w'(l) = w(l)w^*(-l)$, must be real and symmetric, therefore we can consider

48

only half of the weighted coefficients in the calculations. The symmetry properties of the kernel sequence (3.15) can be summarized in the conjugate symmetry property also called Hermitian property:

$$k(l) = k^*(-l) \qquad (3.17).$$

The similar properties enjoyed by the discrete Fourier transform (DFT) allow us to state that the DFT of the sequences are real. If we combine two consecutive kernel sequences:

$$k_{COMB}(l) = k_1(l) + jk_2(l) \qquad (3.18),$$

we can calculate the two distinct DFT by mean of the DFT of $k_{COMB}(l)$ obtaining thus:

$$
\begin{aligned}
DPWVD_1(0, m\Delta f) &= DFT(k_1(l)) \\
&= real(DFT(k_{COMB}(l)))
\end{aligned}
\qquad (3.19),
$$

$$
\begin{aligned}
DPWVD_2(0, m\Delta f) &= DFT(k_2(l)) \\
&= imag(DFT(k_{COMB}(l)))
\end{aligned}
\qquad (3.20).
$$

Therefore combining the two subsequent kernel sequences the number of calculated FFT will be halved.

The property (3.17) allows us, moreover, also to reduce the calculations necessary to evaluate $k_{COMB}(l)$. In fact, we have:

$$
\begin{aligned}
k_{COMB}(l) &= k_1(l) + jk_2(l) \\
k_{COMB}(-l) &= k^*_1(l) + jk^*_2(l)
\end{aligned}
\qquad (3.21),
$$

where l=0,1, .., L. This translates into the ability to calculate the combined kernel sequence using only the positive values corresponding to the positive time of the original sequences, and then to halve the number of complex multiplications needed to compute k'(l).

The second optimized version of the algorithm for calculating the DPWVD that takes into account the symmetry properties of the kernel sequence of Wigner can then be detected in the sequence of operations:

* Evaluation of the analytical signal z(t) from the signal x(t) which length is equal to twice the temporal resolution,

* Calculation of the kernel sequences $k_1(l)$, $k_2(l)$ only for positive values of time,

* Assessment of the combination of the sequences $k_{COMB}(l) = k_1(l) + j k_2(l)$,

* Assessment of the windowed combination $k'_{COMB}(l) = k_{COMB}(l)w(l)$,

* Calculation of the FFT of the sequence $k'_{COMB}(l)$,

* Extraction of the values of DPWVD for 2 consecutive time segments corresponding to the real part and to the imaginary part of FFT.

In Appendix B it is shown the program **PWD.m**, implemented in MATLAB, which executes this optimized algorithm in the case of the input signal is real. However, if the input signal is complex the algorithm itself calculates its PWD but it excludes the routine evaluation for the associated analytical signal.

3.5 ANALYTICAL CALCULATION OF THE SIGNAL

As already noted, the calculation of DPWVD of a real signal x(t), using the analytical signal associated eliminates the concentration of energy around the origin of the frequencies, which is generated by the products between the positive and negative frequencies, and allow the sampling of the signal at the Nyquist frequency.

50

A simple and accurate method for the calculation of the analytical signal z(n)=x(n)+jH[x(n)] is to use its definition in the frequency domain:

$$Z(f) = \begin{cases} 2X(f) , 0 < f < \frac{1}{2} \\ X(f) , f = 0 \\ 0, -\frac{1}{2} < f < 0 \end{cases} \qquad (3.22).$$

From the definition, the analytical signal z(n) can be obtained by calculating the FFT of signal x(n), by canceling the spectrum corresponding to the negative frequencies and then by evaluating his inverse FFT. Assuming that the number of samples of the kernel sequence is N, the number of arithmetic operations needed to calculate the analytical signal is proportional to $N \log_2 N$. In Appendix B is shown the algorithm **HILBERT.m** that executes, in MATLAB, the calculation of the analytical signal.

3.6 EVALUATION OF THE ALGORITHM

From the above, regarding the implementation of an algorithm for the calculation in real-time of the DPWVD, we can deduce two fundamental concepts. The first is that the use of the analytic signal allows us to improve the performance of the DPWVD because it eliminates the problem of aliasing. The second is that by exploiting the symmetry properties of the Wigner kernel sequence, we can calculate the DPWVD using about a half of the calculations necessary for its direct measurement and this result is of the utmost importance in the design of a real time signal processor. The total computational load depends on the following factors:

* Number of samples from a time segment and the other, that is the time resolution,

* Final length of the FFT which determines the frequency resolution,

* Type of arithmetic calculation used in mobile or fixed point, which determines the numerical accuracy of the result.

The upper limit of temporal resolution is just one sample, but often in applications, this resolution may produce redundant information at the cost of an unnecessary computational burden.

The best frequency resolution to use for the DPWVD calculation depends on the particular application. It 'important to note, finally, that while for the STFT a typical length of the FFT of 256 points corresponds to a frequency resolution of 128 points, the DPWVD uses the whole spectrum so that, for a similar frequency resolution, it requires a length of FFT of only 128 points.

The figures below represent the PWVD of some test signals with which it is evaluated the proper functioning of the algorithms described in Appendix B.

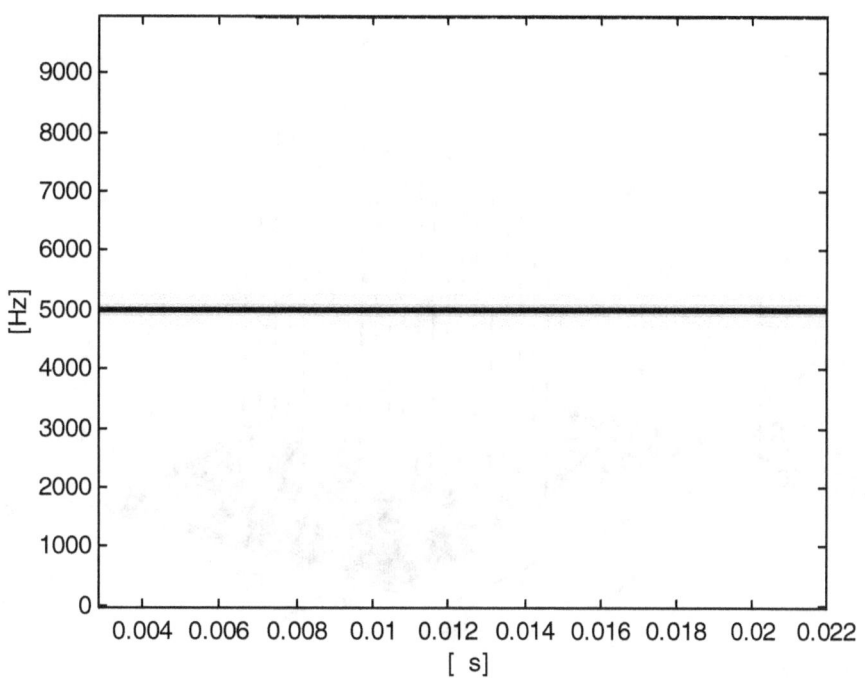

Fig 1.

Fig. 1 shows the pseudo Wigner-Ville distribution of a sinusoidal signal with a frequency equal to 5000 [Hz], sampled at 20000 [Hz], unitary amplitude and duration of 25,6 [ms]. The signal was divided into 49 time segments, whose length is of 128 samples, obtained by performing an overlap of 120 samples.

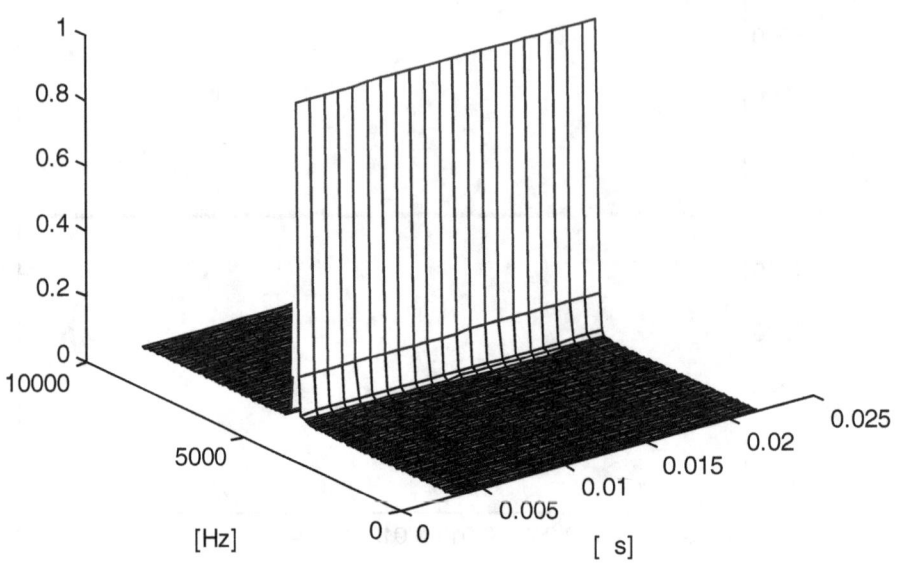

Figure 2.

Fig. 2 shows the three-dimensional representation of the pseudo Wigner-Ville distribution described in Fig. 1.

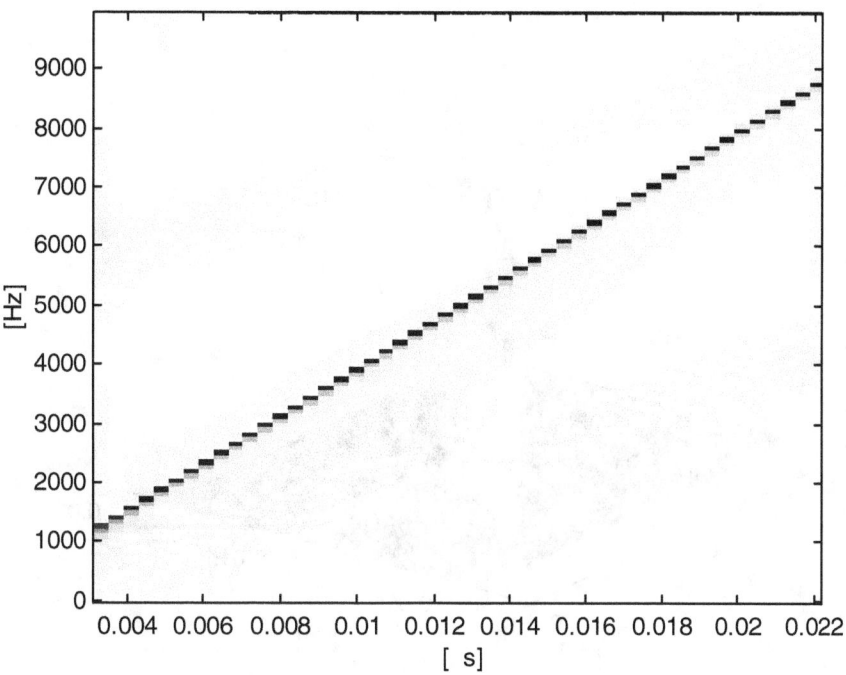

Fig 3.

Fig. 3 shows the pseudo Wigner-Ville distribution of a sinusoidal test signal sampled at 20000 [Hz], with a duration of 25,6 [ms] and whose instantaneous frequency consists of a ramp that goes from 0 [Hz] to 10000 [Hz] (chirp signal). The general form for the instantaneous phase of this signal is:

$$\omega(t) = 2\pi \frac{C}{2} t^2,$$

where C is the slope of the ramp.

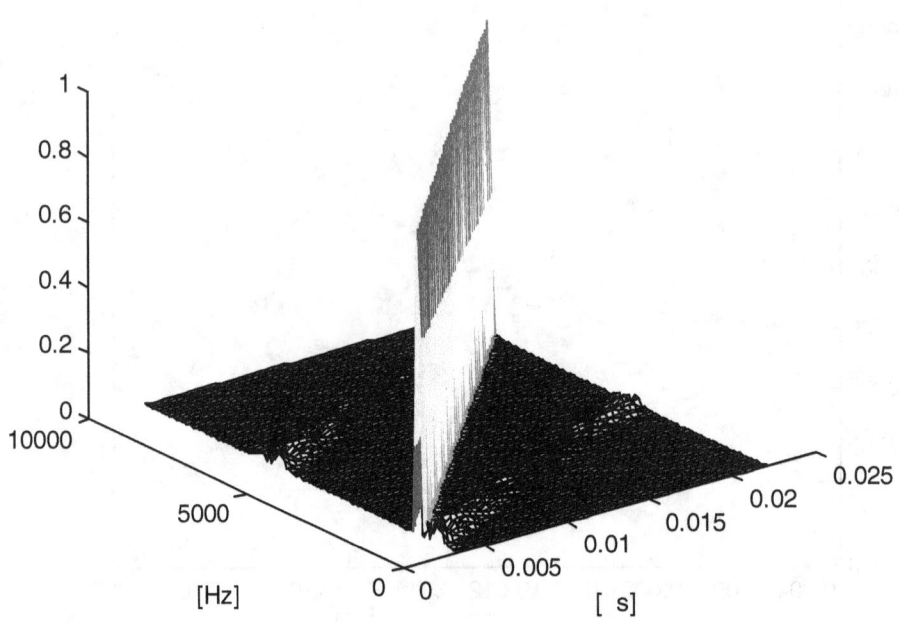

Fig 4.

Fig. 4 shows the three-dimensional representation of the pseudo Wigner-Ville distribution as described in Fig. 3. Note the presence of interference fringes detected, though less obviously, by the gray areas in Figure 3.

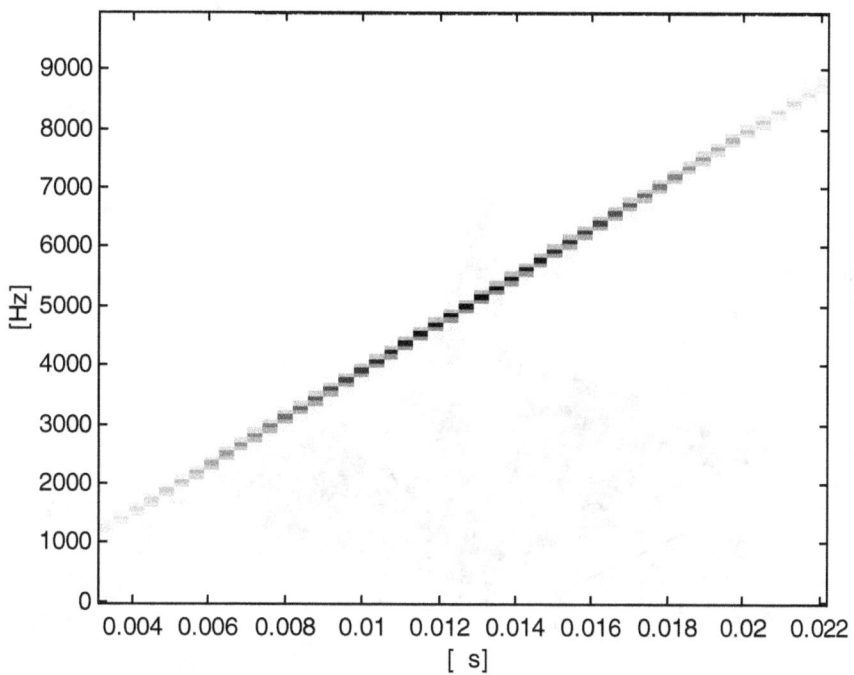

Fig 5.

Fig. 5 shows the pseudo Wigner-Ville distribution of a sinusoidal test signal sampled at 20000 [Hz], with a duration of 25,6 [ms] and whose instantaneous frequency coincides with the signal shown in Fig. 3. In this case, however, the amplitude A(t) is modulated by a sine wave:

$$A(t) = A \cdot sin(2\pi t f_0),$$

where it has been chosen A=10 and f_0=19,5 [Hz].

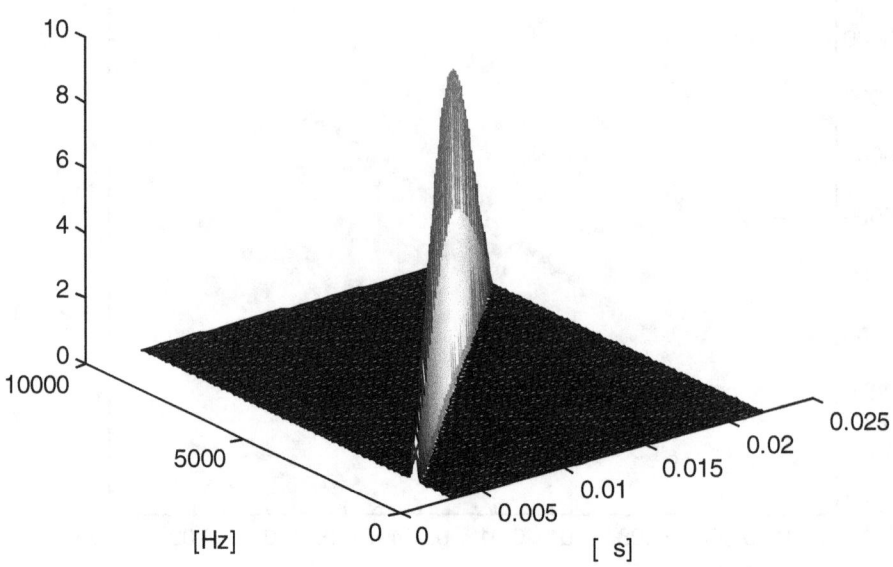

Figure 6.

Fig. 6 shows the three-dimensional representation of the pseudo Wigner-Ville distribution as described in Fig. 5. We can note that in this case there are no interference terms.

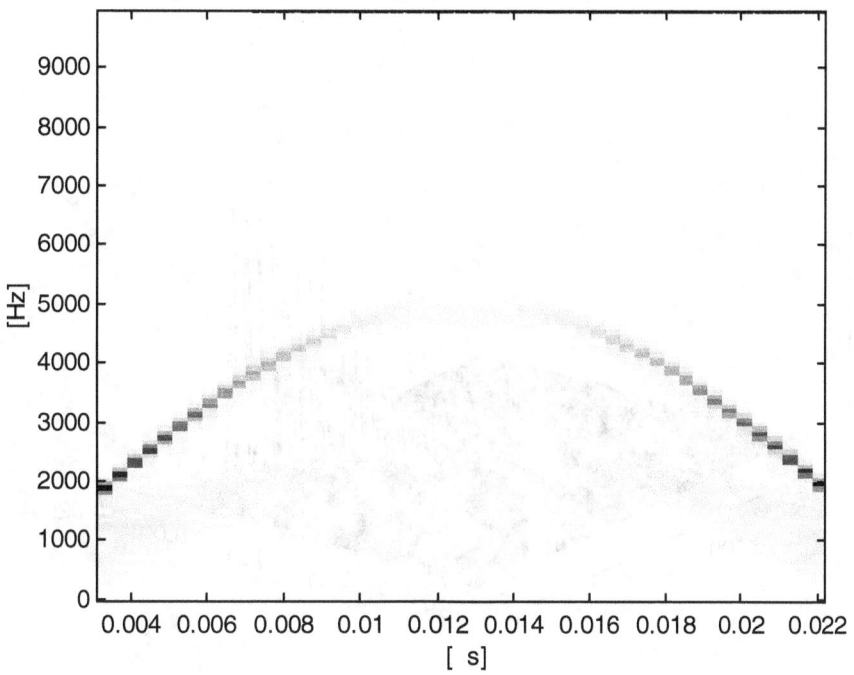

Figure 7.

Fig. 7 shows the pseudo Wigner-Ville distribution of a sinusoidal test signal sampled at 20000 [Hz], with a duration of 25,6 [ms], whose instantaneous pulse $\omega(t)$ is modulated by a sinusoidal signal and whose amplitude A(t) is modulated by a cosine signal.

$$\omega(t) = 2\pi 5000 \cdot sin(2\pi t f')$$

$$A(t) = A \cdot \cos(2\pi t f'),$$

where has been fixed A=10 and f'=19,5 [Hz].

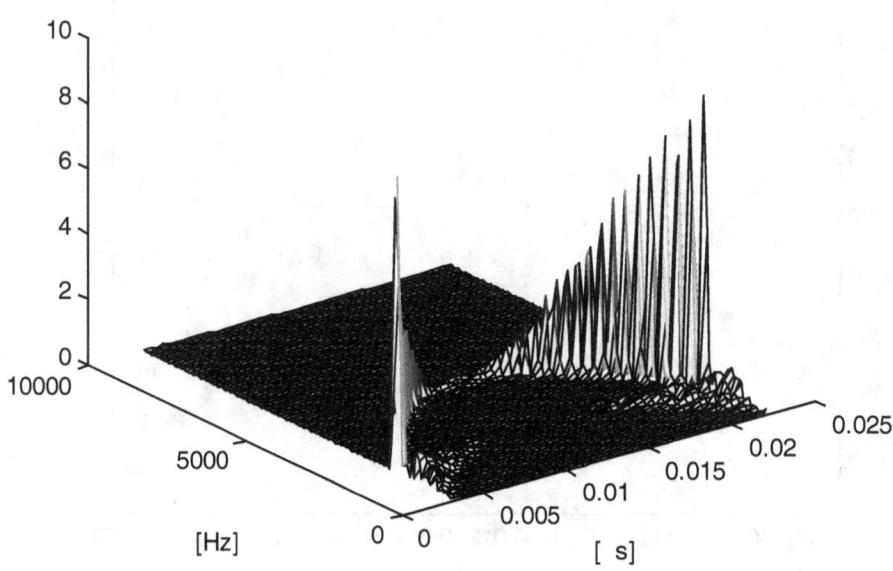

Figure 8.

Fig. 8 shows the three-dimensional representation of the pseudo Wigner-Ville distribution as described in Fig. 7. we can note the presence of interference terms which in fig.7 are shown by grey areas.

CHAPTER 4

ANALYSIS OF THE PULSED DOPPLER SIGNALS BY MEAN OF THE SPECTROGRAM AND OF THE PSEUDO WIGNER DISTRIBUTION

4.1 INTRODUCTION

Many of the current pulsed Doppler ultrasound systems, used for medical diagnosis, adopt the spectrogram as a tool for analysis and visualization of signals broadcasted by the biological tissues. These are non-stationary signals, in fact their characteristics, such as the Doppler spectrum, change over time, so the time-frequency representations (TFR), which describe a one-dimensional signal with a two-dimensional function of time and frequency, are of great help in their process. The TFR as the short Fourier transform (STFT) (1.6) and its corresponding quadratic, the spectrogram (SPEC) (1.17), are appropriate to these signals but have a limited time-frequency resolution in accordance with the principle of indeterminacy. This principle states that the analysis resolution cannot be arbitrarily small because the product of the time resolution Δt and of the frequency resolution Δf is bounded below:

$$\Delta f \Delta t \geq \frac{1}{4\pi} \qquad (4.1).$$

Therefore, there is an inverse relationship between the improvement of frequency resolution and of the temporal resolution. There are other TFR that may be considered as valuable tools for the analysis of non-stationary signals, being equipped with better resolution than the STFT, but there is no an optimal TFR for any signal processing problem. We will turn our attention to the pseudo Wigner distribution (PWD) for which the smoothing function acts only towards the frequency.

In the following we are going to compare the spectrogram and the PWD, whose implementation has been optimized in the form described in the previous chapter, used as tools for the analysis of Doppler ultrasound signals. The performance of the two representations, depending on the Doppler signal bandwidth, is compared in the analysis of echoes from a time variant model of blood flow.

4.2 COMPARISON BETWEEN THE SPECTROGRAM AND THE PSEUDO WIGNER DISTRIBUTION FOR THE ANALYSIS OF PULSED DOPPLER SIGNALS

The *Wigner distribution* (WD) of a continuous signal x(t) which is function of the time, as defined by (3.1), is:

$$WD_x(t, f) = \int_\tau x(t + \tfrac{\tau}{2}) x^*(t - \tfrac{\tau}{2}) e^{-j2\pi f \tau} d\tau \quad (4.2).$$

The WD is a TFR with many desirable mathematical properties, among which there is an excellent time-frequency concentration. The practical application of the WD, however, is often limited by the interference terms due to its quadratic nature. In fact, if we assume that x(t) is a signal with two components:

$$x(t) = y(t) + z(t) \qquad\qquad (4.3),$$

follows that (the mathematical steps are presented in Appendix A):

$$WD_x(t, f) = WD_y(t, f) + WD_z(t, f) +$$

$$+WD_{yz}(t, f) + WD_{zy}(t, f) \qquad (4.4).$$

The first two terms are the auto-WD of y(t) and z(t) and are called *terms of signal* or *auto-terms* of the signal x(t). The other two are the cross-WD of y(t) and of z(t), defined by the (1.21), and are called *interference terms* (IT) or cross terms. For a multi-component signal the number of interference terms increases, according to the relation $\binom{N}{2} = \frac{N(N-1)}{2}$, as the square of the components terms N then the visual analysis of its

WD becomes very difficult. This happen for the ultrasonic echoes produced by the blood motion when the noise components (clutters), coming from surrounding stationary tissues, generate significant IT in the WD of the signal. Unlike the spectrogram, the interference terms of the WD are present even if the auto terms of each component signals do not overlap. This shows that the performance of a time-frequency representation is determined not only by its time-frequency resolution but also by the properties of the interference terms. The latter have an oscillatory nature then can be eliminated or mitigated by a smoothing operation with the result, however, of a reduction in the time-frequency concentration. Here, then, that there is an inverse relationship between a good time-frequency resolution and the presence of small terms of interference. For example, the spectrogram has some valuable properties related to the interference terms, but its time-frequency concentration is low and the opposite is true in the case of the WD.

The limitations of the TFR just seen are overcome by another type of quadratic TFR, already defined in (1.26), the *smoothed Wigner distribution* (SWD):

$$SWD_X(t, f) = WD_X(t, f) * H(t, f) \qquad (4.5),$$

which consists in a 2-D convolution, denoted by the symbol *, in a filtering 2_D, said smoothing function, with the WD of the signal x(t). The SWD is particularly important because it represents the TFR from which derive the spectrogram and the pseudo Wigner distribution. The *spectrogram*, as discussed in (1.17) can be defined as:

$$SPEC_X^W(t, f) = \left| \int_{-\infty}^{+\infty} x(\tau) w^*(\tau - t) e^{-j2\pi f \tau} d\tau \right|^2 \qquad (4.6),$$

where w(t) is the analysis window. As was shown by Classen and Mecklenbrauker [29], we can consider the spectrogram like a special case of the SWD. In fact, if we rewrite the (4.6) in the form of (1.26) we obtain:

$$SPEC_X^W(t, f) = \iint_{t',f'} WD_W(t'-t, f'-f) WD_X(t', f') dt' df' \qquad (4.7),$$

therefore, the spectrogram is the convolution 2-D of the WD of the signal with a smooth function that, not considering the reversal of the axis, turns out to be just the WD of the window w(t):

$$SPEC_X^W(t, f) = WD_X(t, f) * WD_W(-t, -f) \qquad (4.8).$$

The *pseudo Wigner distribution* (PWD) is defined by:

$$PWD(t, f) = \int_{-\infty}^{+\infty} x(t + \tfrac{\tau}{2}) x^*(t - \tfrac{\tau}{2}) h(\tfrac{\tau}{2}) e^{-j2\pi f \tau} d\tau \qquad (4.9),$$

then the PWD may therefore also be seen as a special case of the SWD. In fact, if we consider:

$$PWD_X^h(t, f) = WD_X(t, f) *_w H(f) \qquad (4.10),$$

having denoted with the symbol $*_w$ the 1-D convolution and with H(f) the Fourier transform of h(t), we find the definition (4.5) with H(t,f)=H(f)δd(t).

To clarify the difference between the time-frequency resolution capacity of the PWD and those of the SPEC consider a Gaussian window:

$$h(t) = e^{-t^2/2b} \qquad (4.11),$$

which is used both by the PWD and by the SPEC to treat the signal to be analyzed. The FT of the Gaussian window is:

$$H(\omega) = \sqrt{2\pi b} e^{(-b\omega^2/2)} \qquad (4.12),$$

while its WD is:

$$WD_h(t, \omega) = 2\sqrt{\pi b} e^{(-t^2/b)} e^{-b\omega^2} \qquad (4.13).$$

Thus, from the (4.8), derives that for the Gaussian window, the SPEC is the SWD of the signal with:

$$H(t, \omega) = 2\sqrt{\pi b} e^{(-t^2/b)} e^{-b\omega^2} \qquad 4.14),$$

while for the (4.10), the PWD is the SWD of the signal with:

$$H(t, \omega) = \delta(t) e^{(-b\omega^2/2)} \qquad (4.15).$$

It may be noted that the PWD, in contrast to the SPEC, does not perform any smoothing in the time domain. The expression (4.14) for the SPEC shows the inverse relationship between the temporal resolution and the frequency resolution. In fact, a high value of the parameter b, which determines the length of the window size, produces a low smoothing of the frequency while a strong smoothing in the time dimension and vice versa.

4.3 A STATISTICAL MODEL FOR THE ULTRASOUND ECHOES PRODUCED BY THE BLOOD MOVEMENTS

To compare the performance of the spectrogram and of the pseudo Wigner distribution is necessary to refer to the same type of pulsed Doppler ultrasound signal whose characteristics of interest such as, in our case, the information relating to the speed of blood the flow from which it is produced, are known in advance. In recent years several methods have been used for the evaluation of systems and Doppler diagnostic techniques such as the one that uses a fluid in motion, the one that provides the mechanical movement of a belt, a piston or a wheel, the one that is based on a mathematical model of the tissue flooded by blood. In the method using a fluid with properties similar to the blood this is pumped into a hydraulic circuit suitably made of vinyl tubes of different diameters included in a material that has properties equivalent to those of biological tissues. The mechanical method, however, consist in moving an object which is immersed in a tub filled with aqueous solution similar to the blood and the Doppler shift of the signal is detected by the echoes produced in this way.

To evaluate the performance of the two TFR we will use the signals generated from a simplified model of ultrasound echoes non-stationary produced by the blood movements in which we can vary the parameters related to changes in speed over time, the signal to noise ratio, the amplitude of bandwidth of the Doppler signal.

The general expression of the pattern of ultrasonic echoes is as follows [24]:

$$\tilde{r}(z, t) = \tilde{a}(t)\tilde{s}(z, t - \tau)e^{j\omega_d t} + \tilde{n}(t) \qquad (4.16).$$

The symbol $\tilde{a}(t)$, which indicates the complex envelope of the process $a(t)$ is a complex random process with zero mean, normally distributed, slowly varying and with bandwidth equal to that of the Doppler signal. The time-varying nature of $\tilde{a}(t)$ represents the random behavior of the echoes produced by many transmitter elements in the identified sample volume by the ultrasound beam and situated at a certain depth. Besides $\tilde{n}(t)$ is a normal random process with zero mean and white in the frequency band of interest. $\tilde{s}(z, t)$ is the convolution between $\tilde{f}(t)$, i.e. the output signal from the transducer, and the impulse response of the linear filter $\tilde{h}_{AS}(z, t)$ in which is considered the attenuation effect of the tissues and the diffusion effect by the blood cells on the ultrasonic echoes. Notice that $\tilde{f}(t)$, the output signal from the transducer, is the convolution between the excitation electric signal and the impulse response of the transducer. The symbol τ is the delay corresponding to the depth of the sample volume z, $\tau = \frac{2z}{c}$, where c is the speed of propagation of the ultrasounds in the tissue. Finally $\omega_d = \omega_0 \left(2v_d / c \right)$ is the Doppler shifting where $v_d = v \cos \theta$ is the relation between the velocity of blood v and the angle which is produced between the ultrasound beam and the direction θ of the blood flow.

This model is valid in the cases of transmission of a narrow-band signal and of uniform velocity distribution within the sample volume. We note that assuming v_d constant over the time we obtain a stationary type ultrasound signal, while in reality the speed in a blood vessel can vary considerably during each cardiac cycle. Therefore, the clinical Doppler signal is generally a non-stationary stochastic process.

The model (4.16), in order to take into account the time-varying nature of the blood velocity, can be modified as follows:

$$\tilde{r}(z, t) = \tilde{a}(t)\tilde{s}(z, t - \tau)e^{j\omega_d(t)t} + \tilde{n}(t) \qquad (4.17).$$

In this expression $\omega_d(t)$ is the time variant Doppler frequency shift:

$$\omega_d(t) = \omega_0 \frac{2v_d(t)}{c} \qquad (4.18),$$

where ω_0 is the pulsation of the transmitted pulses by the transducer while $v_d(t)$ is the projection of the velocity vector along the axis of the acoustic transmitter.

In order to concentrate our attention on the problem of non-stationarity of the simulated signals let us omit the effects of frequency-dependent attenuation, of the spread of blood and of the frequency response of the transducer so that can obtain the following simplified model:

$$\tilde{r}(t) = \tilde{a}(t)\tilde{y}(t - \tau)e^{j\omega_d(t)t} + \tilde{n}(t) \qquad (4.19).$$

The function $\tilde{y}(t)$ is the complex envelope of the electrical transmitted signal which happens to be a rectangular pulse with unitary amplitude. $\tilde{a}(t)$ has a variance σ_a^2 and a bandwidth equal to B. We note explicitly that this model assumes that the velocity distribution within the sample volume is uniform. The signal to noise ratio (SNR) input to the TFR is defined as follows:

$$SNR = 10\log_{10}(\sigma_a^2 / \sigma_n^2) \qquad (4.20),$$

where σ_n^2 is the variance of the noise $\tilde{n}(t)$.

Assuming an ideal coherent demodulation of the signal received by the transducer, (4.19) represents the complex signal coming out from the coherent demodulator. The simulation of $\tilde{a}(t)$ has been obtained from:

$$\tilde{a}(t) \;=\; b(t) \;+\; j\,c(t) \qquad\qquad (4.21),$$

where b(t) and c(t) are two random jointly Gaussian processes, mutually uncorrelated, with a zero mean, with variance $\sigma_a^2/2$ and with bandwidth equal to $\left[-(B/2)\,,\,(B/2)\right]$.

Note that to simulate b(t) and c(t) is sufficient to refer to a Gaussian random process, white, with a zero mean and to filter it with a linear filter with a bandwidth equal to B/2. Similarly, we can simulate $\tilde{n}(t)$ given by:

$$\tilde{n}(t) = n_1(t) + jn_2(t) \qquad\qquad (4.22),$$

where $n_1(t)$ and $n_2(t)$ are two jointly Gaussian random processes, mutually uncorrelated, with a zero mean and a variance $\sigma_n^2/2$.

The implementation of the model described has been realized in MATLAB environment and a more detailed description is given in Appendix B.

4.4 NUMERICAL SIMULATION AND DISCUSSION

The comparison between the spectrogram and the pseudo Wigner distribution has been made by running the numerical simulation of Doppler echoes, coherently demodulated, by mean the described model and then comparing the speed estimation error produced by the two TFR algorithms which have been considered.

The simulation of the Doppler signal was carried out for different values of SNR and of Doppler bandwidth amplitude. This was expressed in terms of *bandwidth amplitude normalized* B', where the normalization is calculated with respect to the absolute value of the maximum Doppler shift. So if the maximum Doppler shift is 3000[Hz], a signal with a Doppler bandwidth of 1200[Hz] will have a normalized bandwidth B'=0,4.

The time dependence of the velocity has been modelized by a polynomial of second order:

$$v_d(t) = v_0 + 4(v_m - v_0)\frac{t}{T} - 4(v_m - v_0)\left(\frac{t}{T}\right)^2 \quad (4.23),$$

which well suit to the temporal trends, experimentally observed, of the blood velocity for the cardiac cycles. Typical values of a normal blood flow are T=0,0256 [s], which represents the length of the observation time, v_0=0[m/s] and v_m=0,9[m/s], respectively the minimum and maximum speed.

The transmitted signal is a simple pulse narrow-band carrier frequency f_0=5[MHz] and pulse repetition frequency PRF=20[kHz], typical values of a Doppler system.

Both the two TFR just seen use, for the treatment of the ultrasound signal, the same Gaussian window (4.11):

$$h(t) = e^{-t^2/2b},$$

whose effective length can be defined as: $L_h = 2\sqrt{b}$. To achieve the same smoothing in the frequency dimension for the two TFR, as can be deduced respectively from (4.14) and (4.15), the value of b used for the PWD should be twice that used in the SPEC. Empirical evaluations, designed to obtain the lower estimation frequency error, lead to choose b=800 for the SPEC and b=1600 for the PWD. These values, once set the PRF, correspond to an effective length L_h of the analysis window respectively of 56 and of 80 pulses.

The instantaneous velocity can be estimated from the value of TFR at that moment [25]-[28]. Although the visual analysis of the TFR is often sufficient for the diagnosis, the detection of a procedure for automatic estimation is essential to perform a quantitative analysis of the performance. There are several estimation methods for which the performance depends on the statistical properties of the distribution of velocity and from its time dependence. We will take, for simplicity and to avoid limiting the generality to a particular case, the method which uses the frequency corresponding to the maximum width of the TFR as an estimation of the Doppler shift at any instant.

The calculation of the spectrogram was performed using an existing algorithm properly modified in order to use different analysis windows and to provide an adequate representation of the negative frequencies, or speeds. The algorithm **SPEC.m** has been implemented in MATLAB environment along with the algorithm for calculating the Gaussian window **FINGAUS.m** both quoted in Appendix B. For the calculation of the pseudo Wigner distribution has been used **PWD.m** algorithm, derived from the algorithm **PWVD.m**, described in the previous chapter, which does not perform the preliminary calculation of the analytical signal corresponding to the input signal, if the latter, as in the present case of a Doppler signal coming out from the coherent quadrature demodulator, is a complex signal.

The fig. 1-8 show the images, coded with 64 gray levels, of the ultrasound test signal obtained by using the two transformations for different values of SNR and of normalized Doppler bandwidth. In these figures the horizontal axis corresponds to the time and varies from 0 to 0,0256[s], while the vertical axis, corresponding to frequencies, varies from -5000 to +5000[Hz].

The fig. 1 and 2 show, respectively, the PWD and the SPEC for an high value of the signal to noise ratio, compared to the actual clinical conditions, corresponding to a SNR=20[dB] and to a value of the normalized band B'=0,05, which is also beneficial. It is clearly observed that the image of the speeds obtained with the PWD is much clearer and focused around the real performance compared to that produced by the SPEC.

The fig. 3 and 4 show, respectively, the PWD and the SPEC for SNR=20 dB and for a normalized bandwidth B'=0,4. We note that the traces of the speeds are larger, compared to the ones in Fig. 1 and 2, and in fact correspond to a wider bandwidth of the Doppler signal making the visual comparison between the two representations more difficult.

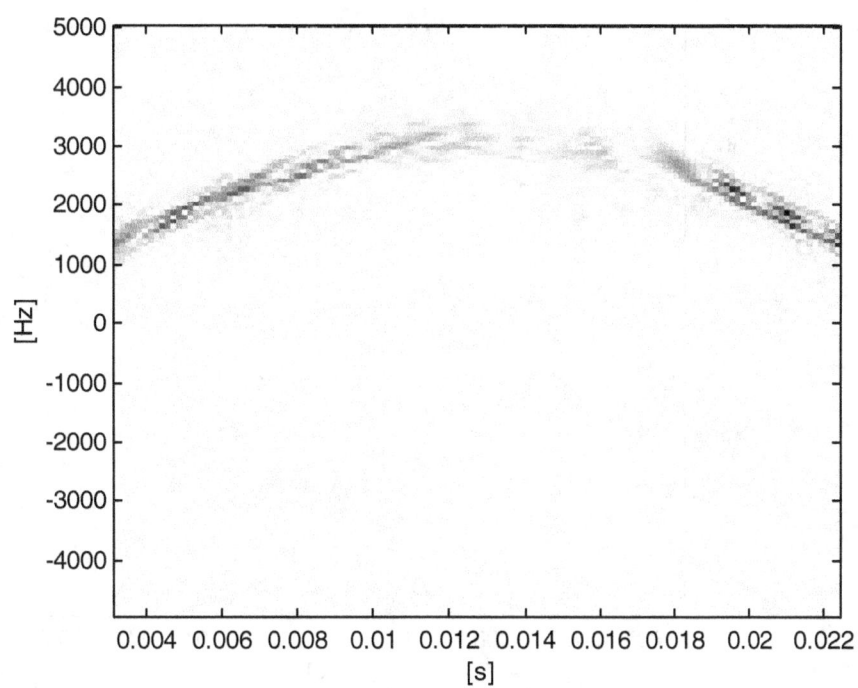

Fig 1. PWD with parameter SNR=20[dB] and B'=0,05.

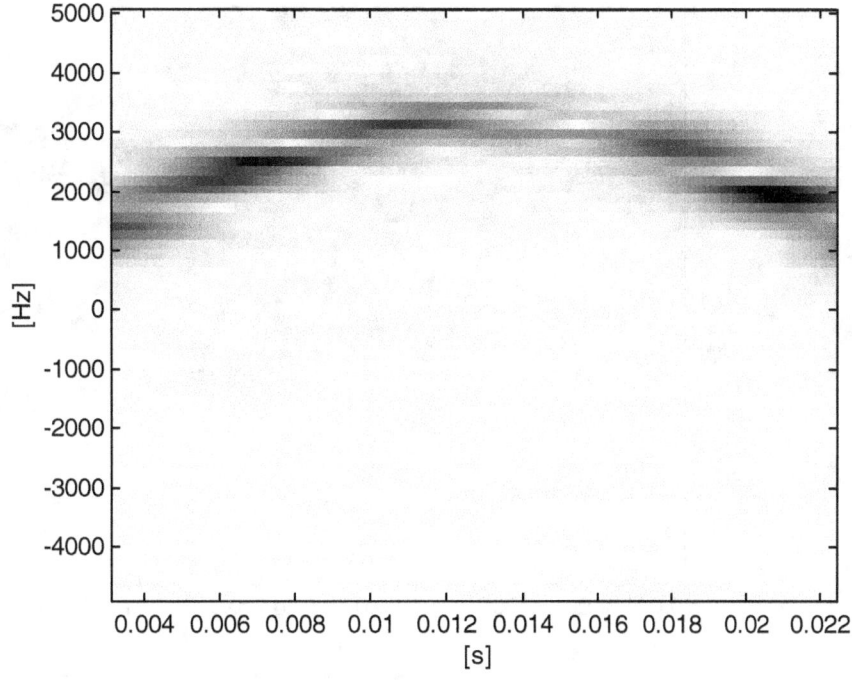

Fig 2. SPEC with parameters SNR=20[dB] and B'= 0,05.

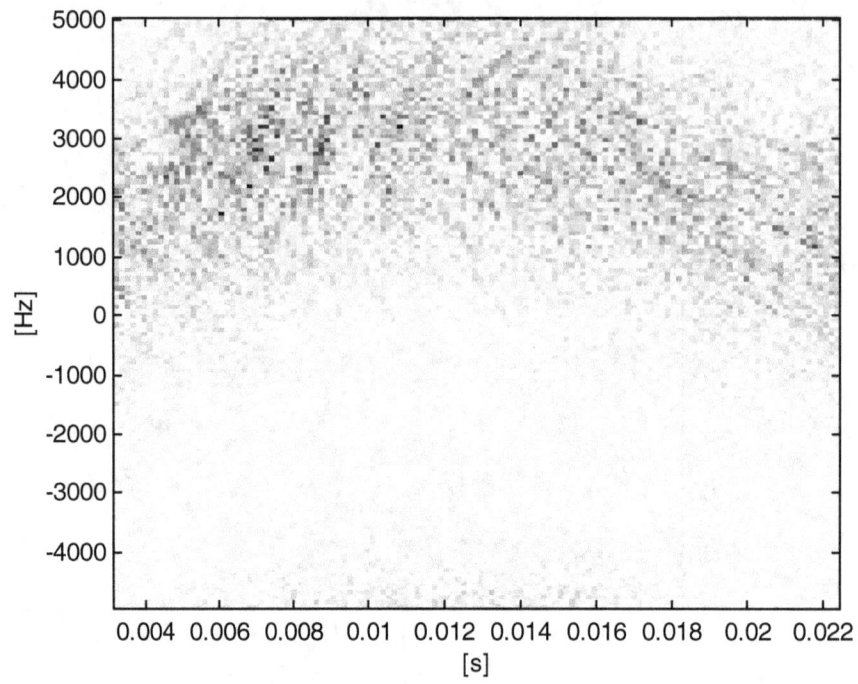

Fig 3. PWD with parameter SNR=20[dB] and B'=0,4.

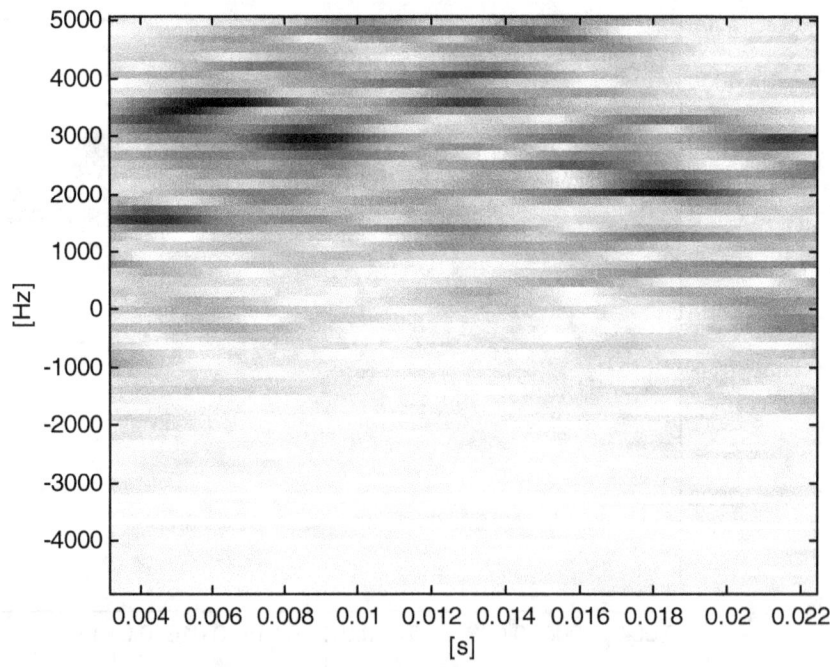

Fig 4. SPEC with parameters SNR=20[dB] and B'=0,4.

The fig. 5 and 6, however, show the PWD and the SPEC corresponding to a SNR=5dB and to a normalized bandwidth B'=0,05. Again, as in the case corresponding to higher values of the SNR, the speeds in the PWD are clearer than the ones in this SPEC. We note, however, that the PWD is characterized by the presence of interference terms or cross terms, between the useful signal and the noise, which produce a background noise (salt-and-pepper) on the entire time-frequency plane. As expected theoretically, this type of background noise appears very slight in the images produced by the SPEC.

The fig. 7 and 8 show the PWD and the SPEC for SNR=5dB and for a normalized bandwidth B'=0,4. Although the parabolic velocity is still discernible in both images, the combination of a low SNR and of a wide Doppler band makes the visual interpretation of the image very uncertain.

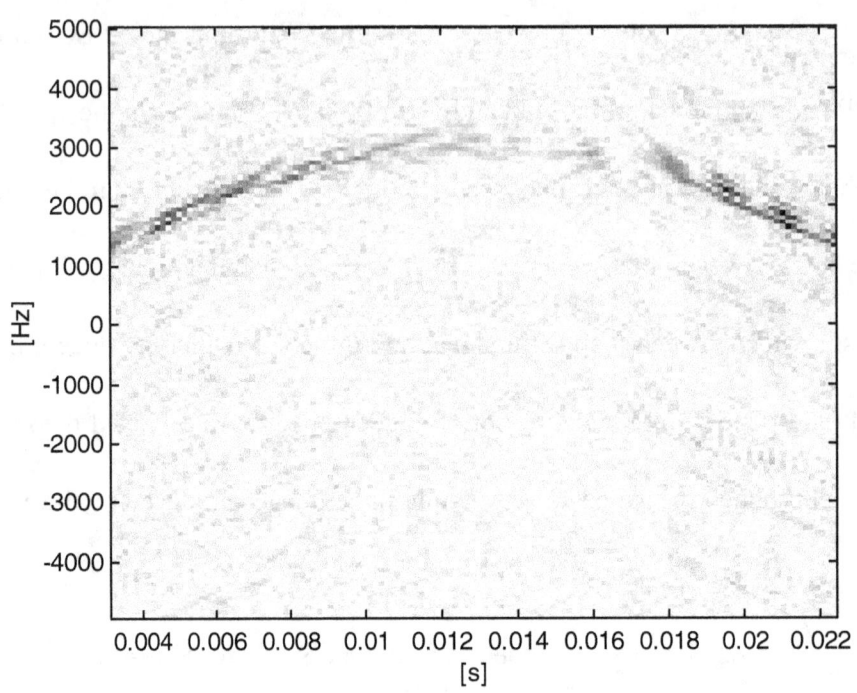

Fig. 5. PWD with parameters SNR=5[dB] and B'=0,05.

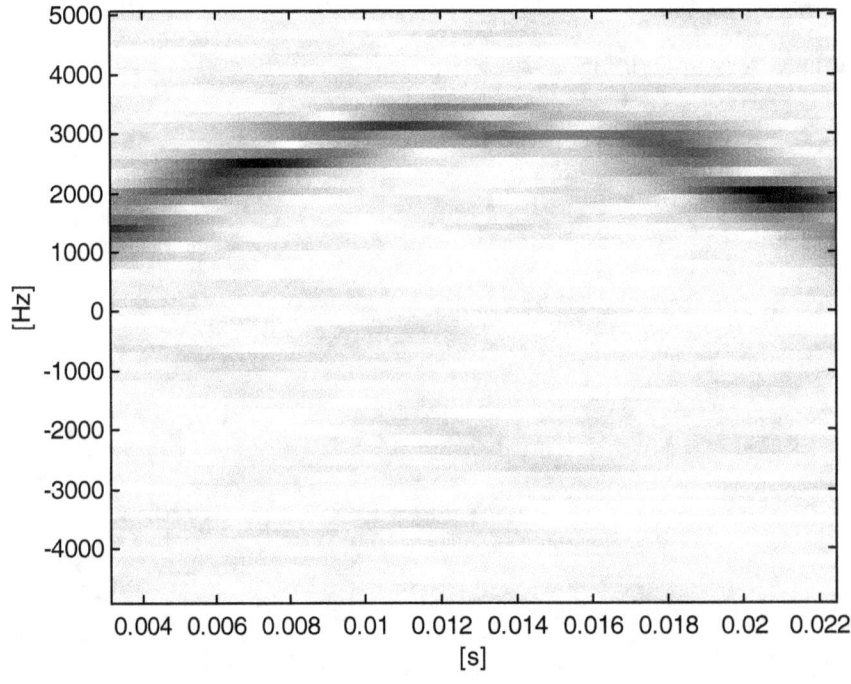

Fig. 6. SPEC with parameters SNR=5[dB] and B'=0.05.

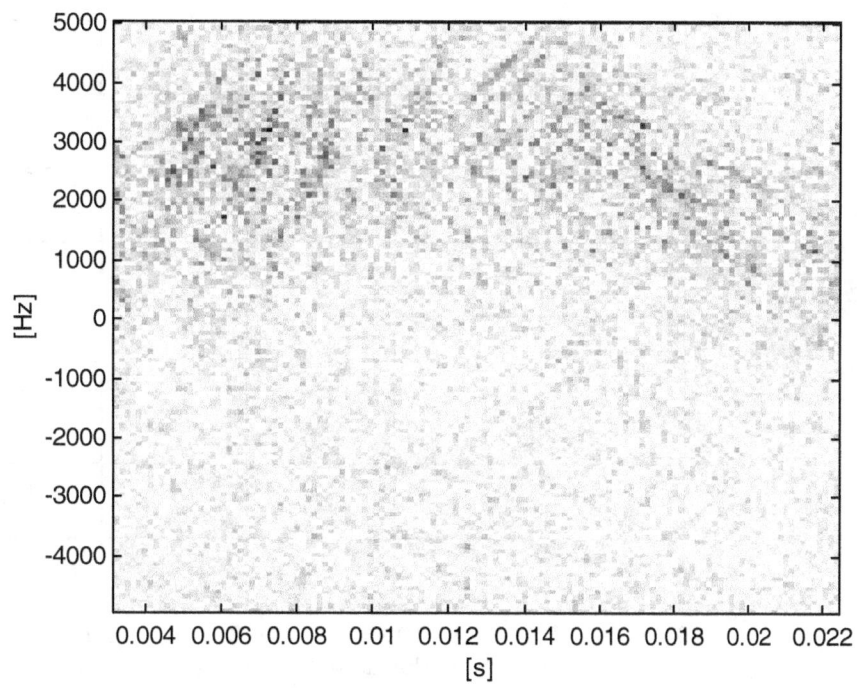

Fig. 7. PWD with parameters SNR=5[dB] and B'=0,4.

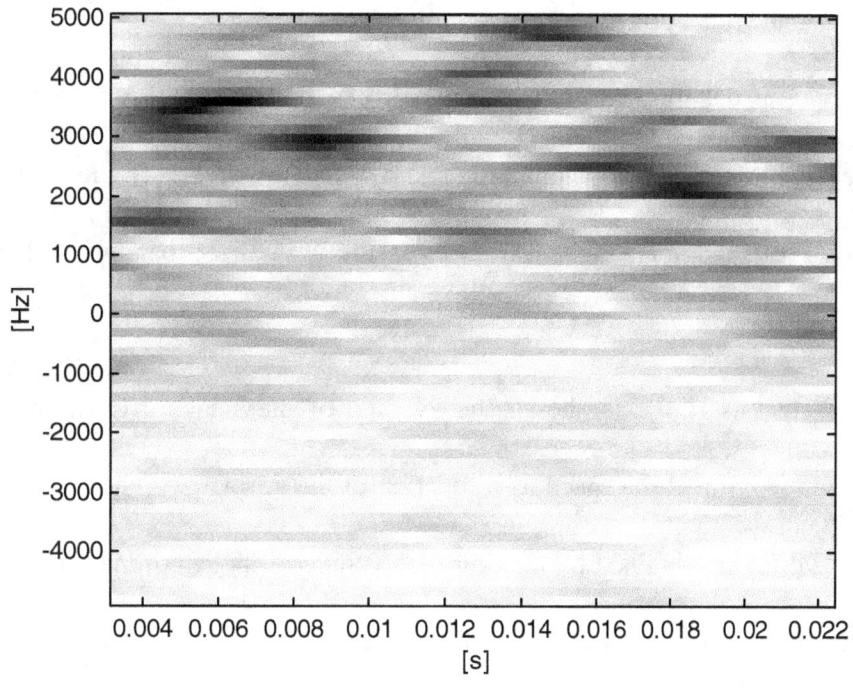

Fig. 8. SPEC with parameters SNR=5[dB] and B'=0,4.

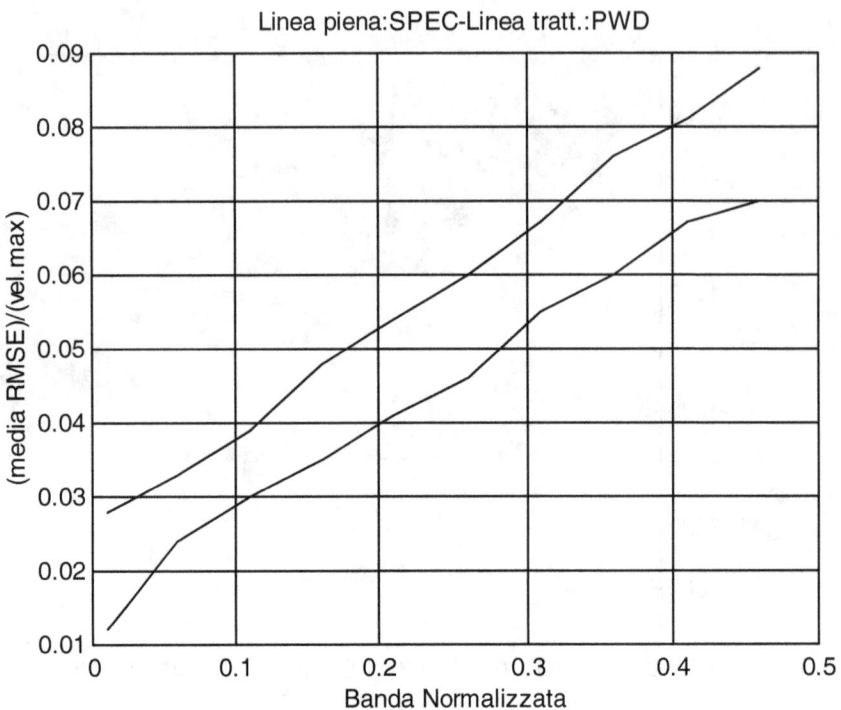

Fig. 9. Average RMSE as a function of B' shown with a solid line (_____) for the

SPEC, and with a dotted line (------) for the PWD.

The quantitative comparison between the two TFR was achieved through an

extensive Monte Carlo simulation. The fig. 9 summarizes the results obtained for

SNR=15dB and for a wide variation of the normalized bandwidth B' of the Doppler

signal.

In this figure is shown the average of the standard deviation (RMSE) of the

estimation error of the speed as a function of the normalized band. The average value of

the RMSE was obtained by making the average of its determinations by the duration of

observation. It is clearly observed that the average value of the RMSE increases linearly

when the normalized Doppler bandwidth increases, and that the RMSE related to the pseudo Wigner distribution, is consistently lower, and then better than that related to the spectrogram.

We conclude the comparison between the two TFR commenting on the values of the computing time used by the implemented algorithms. The computer used has a microprocessor Intel 80386 DX, a math coprocessor Weitek 80387 and a clock frequency of 25 [MHz]. The time measurement, expressed in seconds, was performed using the MATLAB **ETIME.m** function. It was carried out, moreover, the measurement of the floating-point operations number using the algorithm **FLOPS.m**, also provided by the MATLAB environment. We note that the last algorithm does not allow a high accuracy and that it uses, for the real or complex addition and subtraction, respectively, one or two floating point operations. For the real or complex multiplications and divisions, however, it uses respectively one or six floating-point operations.

The algorithms **PWVD.m**, its optimized version **PWD.m** and **SPEC.m**, have been evaluated. within the MATLAB environment, for the analysis of a sine type test real signal:

$$Pr(t)=\sin(2\pi t*2500),$$

and a complex signal:

$$Pc(t)=\cos(2\pi t*2500)+j\sin(2\pi t*2500).$$

These signals have been simulated with a sampling frequency of 20000[Hz], are represented with 512 samples and, therefore, duration of 25,6[ms]. The algorithms have been used with the following values of input parameters: FFT points 128, overlapping

samples 115, Gaussian window FINGAUS.m with 128 points with parameter b=1000.

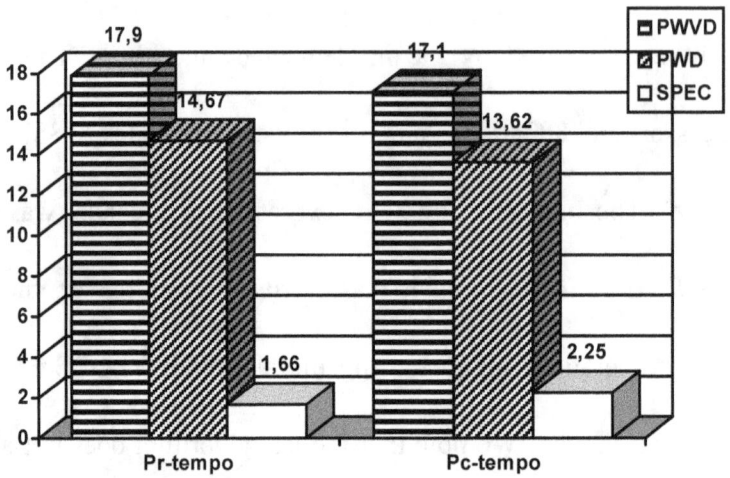

Fig 10. Histogram of the calculation time for the two test signals Pr(t) and Pc(t)

for the three algorithms, expressed in seconds, all having with the same parameters.

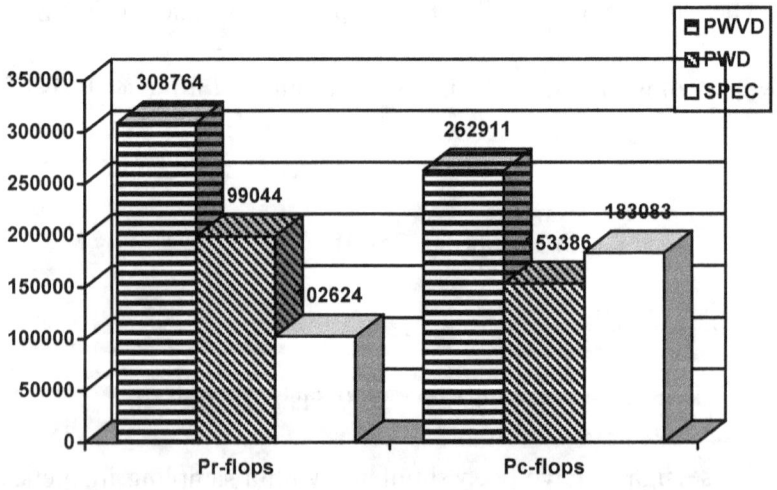

Fig 11. Histogram of the floating-point operations number (FLOPS) performed by

the three algorithms for the two test signals Pr(t) and Pc(t).

From the Fig. 10 we can see clearly, in both cases of real or complex test signal, the substantial reduction of the PWD.m algorithm computation time compared to its non-optimized version PWVD.m, and the clear difference of time calculation between the pseudo Wigner distribution and the spectrogram. This difference is the price that the pseudo Wigner distribution demands, in exchange for better performance offered in certain circumstances, compared to the spectrogram.

The Fig. 11, in the case of real test signal, confirms the trends already seen in Fig. 10 by the number of floating point operations. In the case of a complex test signal, however, it is interesting to note that the number of floating point operations performed by the PWD.m algorithm is less than those made by the algorithm SPEC.m. This fact is justified by recalling that the PWD.m algorithm successfully exploits the symmetry properties of the Wigner kernel sequences that allow it to reduce approximately by half the number of steps than the ones required by PWVD.m algorithm.

CONCLUSIONS

The time-frequency representations of the signals are a powerful tool for the analysis of non-stationary signals such as the biological ultrasound Doppler pulsed type. The measure of their instantaneous frequencies and of the amplitude of their spectral components, which corresponds to the measure of moving tissue velocity, can be performed using two specific types of time-frequency representation: the spectrogram, which is a well-established method of representation and of analysis, and the pseudo Wigner distribution that exhibits better temporal and frequency resolution properties, although the presence of interference terms limits its performance. The two representations, implemented in MATLAB - SIMULINK environment using optimized algorithms which have been listed in the Appendix, are compared analysing a signal simulated from a simple statistical model of Doppler ultrasound echoes coherently demodulated. Comparing the average of the standard deviation of the estimated error of the speed for the two representations, when varying the normalized bandwidth of the Doppler signal and assuming an high value of signal to noise ratio, we can note that the pseudo Wigner distribution exhibit an higher performance than the spectrogram, in agreement with the theoretical properties and with the visual analysis of images. Therefore, the analysis of Doppler ultrasound signals, for medium to high values of signal to noise ratio, can be improved replacing the spectrogram with the pseudo Wigner distribution. The performance of the latter, however, is drastically reduced for low values of signal to noise ratio, due to the presence of interference terms. This negative characteristic leads to investigate other smoothed Wigner distributions, with a greater capacity of noise suppression and for which the smoothing properties can be adjusted according to the characteristics of the signal to be analyzed.

Chapter 1

Below follow the key definitions of the time-frequency representations [1]:

Linear TFR:

Short Time Fourier Transform (STFT) [30], [31], [32]:

$$STFT_X^{(\gamma)}(t,\ f) = \int_{t'}\left[x(t')\gamma^*(t'-t)\right]e^{-j2\pi ft'}dt'$$

$$= e^{-j2\pi ft}\int_{f'}X(f')\Gamma^*(f'-f)e^{j2\pi tf'}df'.$$

Wavelet Transform (WT) [33], [34], [35]:

$$WT_X^{(\gamma)}(t,\ f) = \int_{t'}x(t')\sqrt{\left|f/f_0\right|}\gamma^*\left(\frac{f}{f_0}\ (t'-t)\right)dt'$$

$$= \int_{f'}X(f')\sqrt{\left|f/f_0\right|}\gamma^*\left(\frac{f_0}{f}\ f'\right)e^{j2\pi tf'}df'.$$

Quadratic TFR:

Ambiguity function (AF) :

$$A_{x,y}(\tau,\ \nu) = \int_t x(t + \tfrac{\tau}{2})y^*(t - \tfrac{\tau}{2})e^{-j2\pi\nu t}dt$$

$$= \int_f X(f + \tfrac{\nu}{2})Y^*(f - \tfrac{\nu}{2})e^{j2\pi\tau f}df.$$

Butterworth distribution (BUD) :

$$BUD_X(t,\ f) = \iint_{\tau\nu}\Psi(\tau,\ \nu)A_X(\tau,\ \nu)e^{j2\pi(\nu t - \tau f)}d\tau d\nu,$$

$$\Psi(\tau,\ \nu) = \frac{1}{1+\left(\dfrac{\tau}{\tau_0}\right)^{2M}\left(\dfrac{\nu}{\nu_0}\right)^{2N}}.$$

Exponential Choi-Williams distribution (CWD):

$$CWD_x(t, f) = \iint_{\tau v} \Psi(\tau, v) A_x(\tau, v) e^{j2\pi(vt-\tau f)} d\tau dv ,$$

$$\Psi(\tau, v) = \exp\left[\frac{-(2\pi\tau v)^2}{\sigma}\right].$$

Cone-kernel distribution (CKD) [36]:

$$CKD_x(t, f) = \int_{\tau}\left[\int_{t'}\varphi(t - t', \tau)x(t'+\tfrac{\tau}{2})x^*(t'-\tfrac{\tau}{2})dt'\right]e^{-j2f\tau\pi} d\tau ,$$

$$\varphi(t, \tau) = \begin{cases} g(\tau), & |t/\tau| < 0.5 \\ 0, & |t/\tau| > 0.5 \end{cases}.$$

D-Flandrin distribution (FD) :

$$FD_x(t, f) = f\int_u X\left[f(1 + \tfrac{u}{4})^2\right]X^*\left[f(1 - \tfrac{u}{4})^2\right]\left[1 + (\tfrac{u}{4})^2\right]e^{j2\pi tfu} du$$

Generalized exponential distribution (GED) :

$$GED_x(t, f) = \iint_{\tau v} \Psi(\tau, v) A_x(\tau, v) e^{j2\pi(vt-\tau f)} d\tau dv$$

$$\Psi(\tau, v) = \exp\left[-(\tfrac{\tau}{\tau_0})^{2M}(\tfrac{v}{v_0})^{2N}\right].$$

Generalized Wigner distribution (GWD) :

$$GWD^\alpha{}_x(t, f) = \int_\tau x(t + (0.5 + \alpha)\tau)x^*(t - (0.5 - \alpha)\tau)e^{-j2f\tau\pi} d\tau$$

Pseudo Wigner distribution (PWD) [37], [38], [39]:

$$PWD_x^{(\eta)}(t, f) = \int_{-\infty}^{+\infty} x(t + \tfrac{\tau}{2})x^*(t - \tfrac{\tau}{2})\eta(\tfrac{\tau}{2})\eta^*(-\tfrac{\tau}{2})e^{-j2\pi f\tau} d\tau$$

$$= \int_{-\infty}^{+\infty} H(f - f')WD_x(t, f') df'$$

$$H(f) = \int_{-\infty}^{+\infty} \eta(\tfrac{\tau}{2})\eta^*(-\tfrac{\tau}{2})e^{-j2\pi f\tau} d\tau$$

Rihaczek distribution (RD) :

$$RD_x(t, f) = \int_{-\infty}^{+\infty} x(t + \tau)x^*(t)e^{-j2\pi f\tau} d\tau = x^*(t)X(f)e^{j2\pi ft}$$

Scalogram (SCAL) :

$$SCAL_x^{(\gamma)}(t,\, f) \;=\; \left|WT_x^{(\gamma)}(t,\, f)\right|^2.$$

Smoothed pseudo Wigner distribution (SPWD) :

$$SPWD_x^{(g,\,\eta)}(t,\, f) \;=\; \int_\tau \left[\int_{t'} g(t - t',\, \tau)x(t'+ \tfrac{\tau}{2})x^*(t'- \tfrac{\tau}{2})dt'\right]\eta(\tfrac{\tau}{2})\eta^*(- \tfrac{\tau}{2})e^{-j2f\tau\pi}d\tau$$

$$= \iint_{t'f'} g(t - t')H(f - f')WD_x(t',\, f')dt'\, df',$$

$$H(f) \;=\; \int_{-\infty}^{+\infty} \eta(\tfrac{\tau}{2})\eta^*(- \tfrac{\tau}{2})e^{-j2\pi f\tau}d\tau.$$

Spectrogram (SPEC) :

$$SPEC_x^{(\gamma)}(t,\, f) \;=\; \left|STFT_x^{(\gamma)}(t,\, f)\right|^2$$

Wigner Distribution (WD) :

$$WD_x(t,\, f) \;=\; \int_\tau x(t + \tfrac{\tau}{2})x^*(t - \tfrac{\tau}{2})e^{-j2f\tau\pi}d\tau$$

$$= \int_v X(f + \tfrac{v}{2})X^*(f - \tfrac{v}{2})e^{j2tv\pi}dv.$$

———

Below follow the main mathematical properties that apply to the quadratic time-frequency

representations, with particular reference to the Wigner distribution [1].

1) real function: $WD_x^*(t,\, f) \;=\; WD_x(t,\, f)$

2) time shifts conservation:

$$\tilde{x}(t) \;=\; x(t - t_0) \;\Rightarrow\; WD_{\tilde{x}}(t,\, f) \;=\; WD_x(t - t_0,\, f)$$

3) frequency shifts conservation:

$$\tilde{x}(t) \;=\; x(t)e^{j2\pi f_0 t} \;\Rightarrow\; WD_{\tilde{x}}(t,\, f) \;=\; WD_x(t,\, f - f_0)$$

4) temporal marginal property:

$$\int_f WD_x(t,\, f)df \;=\; p_x(t) \;=\; \left|x(t)\right|^2$$

5) frequency marginal property:

$$\int_t WD_x(t,\ f)dt\ =\ P_x(f)\ =\ |X(f)|^2$$

6) time momentum:

$$\int_t \int_f t^n WD_x(t,\ f)dfdt\ =\ \int_t t^n |x(t)|^2 dt$$

7) frequency momentum:

$$\int_t \int_f f^n WD_x(t,\ f)dfdt\ =\ \int_f f^n |X(f)|^2 df$$

8) time-frequency scaling:

$$\tilde{x}(t)\ =\ \sqrt{|a|}x(at)\ \Rightarrow\ WT_{\tilde{x}}^{(\gamma)}(t,\ f)\ =\ WT_{x}^{(\gamma)}(at,\ \tfrac{f}{a})$$

9) instantaneous frequency:

$$\frac{\int_f fWD_x(t,\ f)df}{\int_f WD_x(t,\ f)df}\ =\ f_x(t)\ =\ \frac{1}{2\pi}\frac{d}{dt}\arg\{x(t)\}$$

10) group delay:

$$\frac{\int_t tWD_x(t,\ f)dt}{\int_t WD_x(t,\ f)dt}\ =\ f_x(f)\ =\ -\frac{1}{2\pi}\frac{d}{df}\arg\{X(f)\}$$

11) finite time support :

$$\text{if x(t)=0 for t} \neq \text{[t1,t2] => } WD_x(t,\ f)\text{=0 for t} \neq \text{[t1,t2]}$$

12) finite frequency support:

$$\text{if X(f)=0 for f} \neq \text{[f1,f2] => } WD_x(t,\ f)\text{=0 for f} \neq \text{[f1,f2]}$$

13) Moyal's formula (unitariety):

$$(WD_{x1,y1},\ WD_{x2,y2})\ =\ (x1,\ x2)\ (y1,\ y2)^*$$

14) convolution :

$$\tilde{x}(t) = \int_{t'} h(t - t')x(t')dt' \Rightarrow$$

$$WD_{\tilde{x}}(t, f) = \int_{t'} WD_h(t - t', f)WD_x(t', f)dt'$$

15) multiplication:

$$\tilde{x}(t) = h(t)x(t) \Rightarrow$$

$$WD_{\tilde{x}}(t, f) = \int_{f'} WD_h(t, f - f')WD_x(t, f')df'$$

16) Fourier transforms:

$$\tilde{x}(t) = \sqrt{|c|}X(ct) \Rightarrow WT_{\tilde{x}}(t, f) = WT_x(-\frac{f}{c}, ct)$$

17) chirp convolution:

$$\tilde{x}(t) = x(t)* \sqrt{|c|}e^{j2\pi \frac{ct^2}{2}} \Rightarrow WT_{\tilde{x}}(t, f) = WT_x(t - \frac{f}{c}, f)$$

18) chirp multiplication:

$$\tilde{x}(t) = x(t)e^{j2\pi \frac{ct^2}{2}} \Rightarrow WT_{\tilde{x}}(t, f) = WT_x(t, f - ct)$$

Chapter 2

With relation to (2.2) and (2.3) follow the below considerations. The average frequency of

the blood is estimated by the frequency spectrum of the ultrasonic echoes. When a sound with

frequency ω_0 is transmitted within the blood, the received echo signal e(t) can be described as:

$$e(t) = Re\left\{z(t)e^{j\omega_0 t}\right\} \qquad (A.2.1),$$

where z(t)=x(t)+jy(t) is the complex envelope of the signal e(t). Using a quadrature phase

demodulator the two components of (A.2.1) can be obtained separately. If we denote the power

spectrum of z(t) with P(ω), then its average frequency is:

$$\overline{\omega} = \frac{\int_{-\infty}^{+\infty} \omega P(\omega)d\omega}{\int_{-\infty}^{+\infty} P(\omega)d\omega} \qquad \text{(A.2.2).}$$

The existence of turbulence in the blood flow can be highlighted by the variance of the spectrum. If we denote the standard deviation of the spectrum with σ, the variance can be calculated from:

$$\sigma^2 = \frac{\int_{-\infty}^{+\infty} (\omega - \varpi)^2 P(\omega)d\omega}{\int_{-\infty}^{+\infty} P(\omega)d\omega} = (\overline{\omega^2}) - (\overline{\omega})^2 \qquad \text{(A.2.3).}$$

An alternative way to calculate the average frequency and the variance is to use the *autocorrelation function* R(τ) for the complex signal z(t) which can be deductible, for the Wiener-Khinchine theorem, by the power spectrum P(ω).

$$R(\tau) = \int_{-\infty}^{+\infty} P(\omega)e^{j\omega\tau} d\omega \qquad \text{(A.2.4).}$$

It follows that:

$$j\overline{\omega} = \frac{\dot{R}(0)}{R(0)}$$

$$\sigma^2 = \left\{\frac{\dot{R}(0)}{R(0)}\right\}^2 - \frac{\ddot{R}(0)}{R(0)} \qquad \text{(A.2.5).}$$

Is there a way, approximate but simpler, to calculate the average frequency and the variance by expressing the autocorrelation function in this way:

$$R(\tau) = |R(\tau)|e^{j\Phi(\tau)} = A(\tau)e^{j\Phi(\tau)} \qquad \text{(A.2.6),}$$

and considering that A(τ) is an even function while $e^{j\Phi(\tau)}$ is an odd function. From the calculation of the first derivative of R(t) we have:

$$\dot{R}(\tau) = (\dot{A}(\tau) + jA(\tau)\dot{\Phi}(\tau))e^{j\Phi(\tau)} \qquad \text{(A.2.7),}$$

therefore:

$$\dot{R}(0) = jA(0)\dot{\Phi}(0)$$

$$R(0) = A(0) \qquad (A.2.8).$$

Thus follows:

$$\varpi = \dot{\Phi}(0) \cong \{\Phi(T) - \Phi(0)\}/T = \Phi(T)/T \qquad (A.2.9),$$

where T is the emission interval of the ultrasonic pulses, during which we assume, approximately,

that the cell blood movement to be constant. If we make the derivative of the (A.2.7) and evaluate

it at $\tau = 0$ we have:

$$\ddot{R}(0) = -A(0)(\dot{\Phi}(0))^2 + \ddot{A}(0) \qquad (A.2.10),$$

where:

$$\sigma^2 = -\ddot{A}(0)/A(0) \qquad (A.2.11).$$

Expanding in series A(τ) around τ and considering that it is an even function we have:

$$A(\tau) = A(0) + \frac{\tau^2}{2}\ddot{A}(0) + \dots \qquad (A.2.12).$$

Neglecting the terms from the third order up, we finally obtain that:

$$\sigma^2 \cong \frac{2}{T^2}\left\{1 - \frac{A(\tau)}{A(0)}\right\} = \frac{2}{T^2}\left\{1 - \frac{|R(T)|}{R(0)}\right\} \qquad (A.2.13).$$

In relation to the spectrum of fig. 11, the function which generates it, is obtained from the

following mathematical steps:

$$X(f) = \int_{-\infty}^{+\infty} x(t)e^{-j2\pi ft}dt = \int_{-\infty}^{+\infty} (x_R(t) + jx_I(t))e^{-j2\pi ft}dt$$

$$X(f) = \int_{-\infty}^{+\infty} A\cos(2\pi f_a t)e^{-j2\pi ft}dt + \int_{-\infty}^{+\infty} B\sin(2\pi f_a t)e^{-j2\pi ft}dt$$

$$+j\int_{-\infty}^{+\infty} A\sin(2\pi f_a t)e^{-j2\pi ft}dt \ +j\int_{-\infty}^{+\infty} B\cos(2\pi f_a t)e^{-j2\pi ft}dt.$$

$$X(f) = \frac{A}{2}\,\delta(f-f_a) + \frac{A}{2}\,\delta(f+f_a) - j\frac{B}{2}\,\delta(f-f_b)$$

$$+j\frac{B}{2}\,\delta(f+f_b) + \frac{A}{2}\,\delta(f-f_a) - \frac{A}{2}\,\delta(f+f_a)$$

$$+j\frac{B}{2}\,\delta(f-f_b) + \frac{B}{2}\,\delta(f+f_b),$$

then you get:

$$X(f) = A\delta(f-f_a) + jB\delta(f+f_b).$$

Chapter 4

In the expression (4.4) we have:

$$WD_x(t,\ f) = \int_\tau [y(t+\tfrac{\tau}{2})z(t+\tfrac{\tau}{2})]\cdot[y(t-\tfrac{\tau}{2})z(t-\tfrac{\tau}{2})]^*e^{-j2\pi f\tau}d\tau$$

$$= \int_\tau y(t+\tfrac{\tau}{2})y^*(t-\tfrac{\tau}{2})e^{-j2\pi f\tau}d\tau \ + \int_\tau z(t+\tfrac{\tau}{2})z^*(t-\tfrac{\tau}{2})e^{-j2\pi f\tau}d\tau$$

$$+\int_\tau y(t+\tfrac{\tau}{2})z^*(t-\tfrac{\tau}{2})e^{-j2\pi f\tau}d\tau \ + \int_\tau z(t+\tfrac{\tau}{2})y^*(t-\tfrac{\tau}{2})e^{-j2\pi f\tau}d\tau$$

$$= WD_y(t,\ f) + WD_z(t,\ f) + WD_{yz}(t,\ f) + WD_{zy}(t,\ f).$$

The first two terms are the auto-WD of Y(t) and of Z(t) and are defined as *terms of signal* or *auto-*

terms of the signal X(t). The other two are the cross-WD of (y(t),z(t)) and of (z(t),y(t))

respectively, defined by:

$$WD_{yz}(t,\ f) = \int_\tau y(t+\tfrac{\tau}{2})z^*(t-\tfrac{\tau}{2})e^{-j2\pi f\tau}d\tau,$$

and are called *interference terms* (IT) or cross terms.

APPENDIX B

These algorithms have been implemented in MATLAB environment (classroom version

4.2) with the help of the SIMULINK program (version 1.3) and of the functions contained in the

Signal Processing TOOLBOX (version 2.0).


```
==================================================
```

Function Yo=PWVD(x, nfft, noverlap, win)
```
%
% The function B=PWVD(X, NFFT, NOVERLAP, WIN) calculates the pseudo
% Wigner-Ville distribution of a real signal contained in the vector X.
% The signal is processed into its analytical form by mean of
% the Hilbert function, then is divided into segments that can
% be overlapped by a number of samples determined by
% the parameter NOVERLAP.
% Each segment is used for the calculation of the Wigner-Ville kernel sequence
% by mean of an analysis window called WINDOW (nfft)
% whose length must match with the parameter NFFT. Then is executed the
% Fourier transform on NFFT points for every column of the matrix B
% corresponding to a fixed time instant.
% The frequencial values vary along the rows of the matrix B and
% are ascending ordered, starting with the negative components, down,
% so that the continous component remains in the middle of the representation.
% If NX is the length of the vector X, B is a complex matrix with NFFT
% rows and FIX((NX-NOVERLAP)/(NFFT-NOVERLAP)) columns.
%
% University of Naples "Federico II" - Faculty of Engineering.
% Thesis: "The Pseudo Wigner Distribution for the Analysis of Doppler
% Ultrasound Signals"
%
% Supervisor: Prof. Felice Cennamo.
% Candidate: Dario Fresa, matr.:15/15049.
% Academic Year 1993/1994.
% Graduation session 28/2/1995.
%
% Calculation of the analytical signal for x using the HILBERT function.
x=hilbert(x);
% Calculation of the size of the matrix y.
nx=length(x);
ncol=fix((nx-noverlap)/(nfft-noverlap));
y=zeros(nfft,ncol);
% Now I put the vector x in the columns of y with the appropriate row and column
```

91

```
% index after having transformed the row vector x in a column.
x=x(:);
colIndex=1+(0:(ncol-1))*(nfft-noverlap);
rowIndex=(1:nfft)';
y(:)=x(rowIndex(:,ones(1,ncol))+colIndex(ones(nfft,1),:)-1);
t=nfft;
sec=win.*conj(win);
        % Compute cycles of the KERNEL segments
        % Take a segment of the signal and we set it in 'k' properly.
        for i=1:1:ncol,
                k=y(:,i);
                for h=nfft:-1:(nfft/2+1),
                k(h)=k(h-1);
        end;
        for h=1:1:t/2,
                k1(h)=k(t/2+h)*conj(k(t/2+1-h))*sec(t+h/2);
        end;
        k1(t/2+1)=0;
        for h=t/2+2:1:t,
                k1 (h) = k (ht/2-1) * conj (k (-h + t + t / 2 +2)) * sec (ht/2-1);
        end;
        % Let's put within the column i-th of the matrix 'z' the sequence
        % of the i-th segment
        z((t:-1:1),i)=2*k1';
        end;
        % End of calculation of the KERNEL segments.
% Calculation of the fast Fourier transform matrix z.
z=fft(z);
% This procedure shifts the spectrum in the middle of the representation.
for i=1:ncol;
        z(:,i)=fftshift(z(:,i));
end;
% Use of the MATLAB function 'imagesc' to represent the result
% if the function PWVD is not assigned to any variable.
If nargout=0
        imagesc(z);axis xy
else
        Yo=z;
end ================================================
```

===

Function Yo=PWD(x, nfft, noverlap, win)

```
%
% The function B=PWD(X, NFFT, NOVERLAP, WIN) calculates the pseudo
% Wigner distribution of a real or complex signal contained
% within a vector X by mean of an optimized algorithm.
% The signal, if it is real, is transformed in its
% analytic version by mean of the HILBERT function, then it is divided into
% segments that can be overlapped by a number of samples
% which are determined by the parameter NOVERLAP.
% Each segment is used for the calculation of the Wigner kernel sequence
% by mean of an analysis window called WINDOW (nfft)
% whose length must match with NFFT.
% This optimized version of PWD calculation exploits the symmetry properties
% of the kernel sequences. In fact, is considered the combination
% of two segments each time for the calculation of their kernel sequences.
% Their symmetry provides to halve the number of complex multiplications
% required for the calculation of the sequences and of the number of Fourier
% transform over the NFFT points. In this case too each column of the
% matrix B correspond to a fixed instant in time.
% The frequencial values vary along the rows of the matrix B and
% are ascending ordered, starting with the negative components, down,
% so that the continous component remains in the middle of the representation.
% If NX is the length of the vector X, B is a complex matrix with NFFT
% rows and FIX((NX-NOVERLAP)/(NFFT-NOVERLAP)) columns.
%
% University of Naples "Federico II" - Faculty of Engineering.
% Thesis: "The Pseudo Wigner Distribution for the Analysis of Doppler
% Ultrasound Signals"
%
% Supervisor: Prof. Felice Cennamo.
% Candidate: Dario Fresa, matr.:15/15049.
% Academic Year 1993/1994.
% Graduation session 28/2/1995.
%
% If the signal x is real then let's calculate its analytical signal.
if ~ any (any (imag (x)))
        x = hilbert (x);
end
% Calculation of the size of the matrix y.
nx = length (x);
ncol = fix ((nx-noverlap) / (nfft-noverlap));
y = zeros (nfft, ncol);
% Now I put the vector x in the columns of y with the appropriate row and column
% index after having transformed the row vector x in a column.
x = x (:);
colIndex + = 1 (0: (ncol-1)), * (nfft noverlap-);
rowIndex = (1: nfft) ';
y (:) = x (rowIndex (:, ones (1, ncol)) + colIndex (ones (nfft, 1 ),:)- 1);
t = nfft;
```

93

```
sec = win .* conj (win);
% Parity control of the number of columns
if rem (ncol, 2) % ncol is odd
        ncoll = ncol-1;
else % ncol is even
        ncoll = ncol;
end;
        % Compute cycles of the KERNEL segments
        % Take two segments of the signal and set it in 'k1' and in 'k2' properly.
        for i = 1:2: ncoll,
                k1 = y (:, i);
                for h = t: -1: (t / 2 +1),
                        k1 (h) = k1 (h-1);
                end;
                k2 = y (:, (i +1));
                for h = t: -1: (t / 2 +1),
                        k2 (h) = k2 (h-1);
                end;
                for h = 1:1: t / 2,
                        KB1 (h) = k1 (t / 2 + h) * conj (k1 (t / 2 +1- h)) * sec (t + h / 2);
                end;
                KB1 (t / 2 +1) = 0;
                for h = t: -1: (t / 2 +2),
                        KB1 (h) = conj (KB1 (t +2- h));
                end;
                for h = 1:1: t / 2,
                        KB2 (h) = k2 (t / 2 + h) * conj (k2 (t / 2 +1- H)) * sec (t + h / 2);
                end;
                KB2 (t / 2 +1) = 0;
                        for h = t: -1: (t / 2 +2),
                KB2 (h) = conj (KB2 (t +2- h));
        end;
                tras = fft (2 * (* KB2 KB1 + j));
                z (:, i) = real (tras) ';
                z (:, (i +1)) = imag (tras) ';
                end;
                % End of calculation of the segments KERNEL.
% This procedure translates the spectrum in the middle of the representation.
for i = 1: ncoll;
        z (:, i) = fftshift (z (:, i));
end;
% Use of MATLAB function 'imagesc' to represent the result
% if the function PWD is not assigned to any variable.
nargout if == 0
        imagesc (z) axis xy
else
y = z;
end
```

==

```
============================================================
```
function w= FINGAUS(n,b)

%

% The function FINGAUS (N,b) returns the Gaussian window with length N and

% Effective length L=2*sqrt(b).

%

% University of Naples "Federico II" - Faculty of Engineering.

% Thesis: "The Pseudo Wigner Distribution for the Analysis of Doppler

% Ultrasound Signals"

%

% Supervisor: Prof. Felice Cennamo.

% Candidate: Dario Fresa, matr.:15/15049.

% Academic Year 1993/1994.

% Graduation session 28/2/1995.

%

if rem (n,2)

% Window length is odd.

```
        t = n-1;
        w = exp (- (-t / 2:1: t / 2 t )'.*(- / 2:1: t / 2) '/ (2 * b));
```
else

```
        % The length is even.
        w1 = exp (- (-n / 2:1: -1 )'.*(- n / 2:1: -1) '/ (2 * b));
        w2 = exp (- (1:1: n / 2 )'.*( 1:1: n / 2) '/ (2 * b));
        w = [w1 'w2'] ';
```
end

```
============================================================
```

==

function Y0 = SPEC(x, nfft, noverlap, window)
%
% The function B= SPEC(X, NFFT, NOVERLAP, WINDOW) calculates the
% Spectrogram of a real or complex signal in the vector X.
% The signal is divided into segments that can be overlapped
% by a number of samples determined by the parameter NOVERLAP
% Each segment is multiplied by an analysis window WINDOW (nfft)
% whose length must match with the parameter NFFT. Then is executed the
% Fourier transform on NFFT points which produces the estimation of the
% spectrum contained within the signal segments.
% The frequency varies across the rows of the matrix B. If X is a real signal
% are represented only the first NFFT / 2 rows that match,
% from bottom to top, at frequencies from 0 to the Nyquist frequency.
% If X is a complex signal are represented in NFFT rows
% so as to put the frequencies in ascending order by-freq. Nyquist,
% down, to + freq. Nyquist, top, and the continous component in the middle of
% the representation.
% If NX is the length of the vector X, B is a complex matrix with NFFT
% rows and FIX ((NX-NOVERLAP) / (NFFT-NOVERLAP)) columns.
% The function SPEC(...) represents the absolute value of the matrix B,
% using the functions IMAGESC (B) and AXIS XY.
%
% L. Shure 01/01/1991
% Copyright (c) 1991 by The MathWorks, Inc.
%
% University of Naples "Federico II" - Faculty of Engineering.
% Thesis: "The Pseudo Wigner Distribution for the Analysis of Doppler
% Ultrasound Signals"
%
% Supervisor: Prof. Felice Cennamo.
% Candidate: Dario Fresa, matr.:15/15049.
% Academic Year 1993/1994.
% Graduation session 28/2/1995.
%
% Calculation of the size of the matrix y.
nx = length (x);
ncol = fix ((nx-noverlap) / (nfft-noverlap));
y = zeros (nfft, ncol);
% Now I put the vector x in the columns of y with the appropriate row and column
% index after having transformed the row vector x in a column.
colIndex = 1 + (0: (ncol-1)) * (nfft-noverlap);
rowIndex = (1: nfft) ';
x = x (:);
y (:) = x (rowIndex (:, ones (1, ncol)) + colIndex (ones (nfft, 1),:)- 1);
% Applies the selected window to the segments of the signal.
y (:) = window (:, ones (1, ncol)) .* y;
% Performs the fast Fourier transform for each column of y.
y = fft (y);

% If x is real will be represented only the first NFFT / 2 lines corresponding
% to the positive semi-spectrum, if x is complex then will use all lines of y
% shifting them so to move to the center of the continuous spectral component.
if ~ any (any (imag (x)))
        yy = y (1: nfft / 2,:);
else
% This procedure translates the spectrum in the middle of the representation.
for i = 1: ncol;
        y (:, i) = fftshift (y (:, i));
end;
end

% Use of MATLAB function 'imagesc' for y absolute value to represent the result.
nargout if == 0
        imagesc (abs (yy)), axis xy
else
y = yy;
end
==

```
==================================================
```
function y= HILBERT(x)

%

% HILBERT (X) performs the calculation of the analytical signal of X

% The real part of the result is the same signal, while

% the imaginary part is the Hilbert transform of the signal.

%

% If X is a matrix signal, HILBERT (X) transforms the columns

% of X independently.

% Charles R. Denham, January 7, 1988.

% Revised by LS, 11-19-88, 5-22-90, TPK, 11/04/1992

% Copyright (C) 1988, 1990, 1992 The MathWorks, Inc.

% Reference: Jon Claerbout, Introduction to

% Geophysical Data Analysis.

%

% University of Naples "Federico II" - Faculty of Engineering.

% Thesis: "The Pseudo Wigner Distribution for the Analysis of Doppler

% Ultrasound Signals"

%

% Supervisor: Prof. Felice Cennamo.

% Candidate: Dario Fresa, matr.:15/15049.

% Academic Year 1993/1994.

% Graduation session 28/2/1995.

[r, c] = size (x);

if r == 1

        x = x.'; % x is put in the column form

end;

[n, cc] = size (x);

m = 2^ nextpow2(n);

% Calculation of the analytical signal of x using its definition in

% the frequency domain.

y = fft (real (x), m);

if m ~ = 1

        % Creates a vector for the modification of the spectrum Y of x erasing

        % the negative frequency components.

        h = [1, 2 * ones (fix ((m-1) / 2), 1), 1, zeros (fix ((m-1) / 2), 1)];

        y (:) = y * h (:, ones (1, cc));

end

% Performs the inverse Fourier transform in order to obtain a signal

% within the time domain.

y = IFFT (y, m);

y = y (1: n,:);

if r == 1

        y = y. ';

end
```
```

The simplified model of the non-stationary ultrasound echoes produced by the blood motion (4.19), where you can change the parameters related to speed varying over time, the signal to noise ratio, the bandwidth of the Doppler signal,

$$\tilde{r}(t) = \tilde{a}(t)\tilde{y}(t - \tau)e^{j\omega_d(t)t} + \tilde{n}(t) \quad (4.19)$$

has been implemented in MATLAB using the functions contained in the SIMULINK sub-program.

Recall that $\tilde{y}(t)$ is the complex envelope of the electrical signal transmitted which we suppose to be a rectangular pulse with an unity amplitude. $\tilde{a}(t)$ is a complex Gaussian process with variance σ_a^2 and bandwidth equal to B. The signal to noise ratio (SNR), input to the TFR, is defined as follows:

$$SNR = 10 \log_{10}(\sigma_a^2 / \sigma_n^2) \quad (4.20),$$

where σ_n^2 is the variance of the noise $\tilde{n}(t)$.

Assuming an ideal coherent demodulation of the signal received by the transducer, the (4.19) represents the complex signal in output from the coherent demodulator. The simulation of $\tilde{a}(t)$ was obtained from:

$$\tilde{a}(t) = b(t) + jc(t) \quad (4.21),$$

where b(t) and c(t) are two jointly Gaussian random processes, mutually uncorrelated, with a zero mean, variance $\sigma_a^2 / 2$ and bandwidth equal to $\left[-(B/2) , (B/2)\right]$. We note that to simulate b(t) and c(t) it is sufficient to refer to a Gaussian stochastic process, white, with zero mean and filtering it with a linear filter with bandwidth equal to B/2. Similarly, we can simulate $\tilde{n}$ (t)

given by:

$$\tilde{n}(t) \ = \ n_1(t) \ + \ jn_2(t) \ (4.22),$$

where n_1 (t) and n_2 (t) are two jointly Gaussian random processes, mutually uncorrelated,

with zero mean and variance $\sigma_n^2/2$.

The signals n_1 (t) and n_2 (t) were generated using the function, pre defined within MATLAB,

RANDN.m using some values of the pseudo-noise generation seed different so as to guarantee

their to be mutually uncorrelated. For the signals b(t) and c(t) was done similarly by adding a

low-pass filtering in order to obtain the desired bandwidth of the process. The modulating signal

$\omega_d(t)$ was simulated by combining math blocks contained within SIMULINK.

BIBLIOGRAPHY

Chapter 1

[1] F.Hlawatsch and G.F.Boudreaux-Bartels,"Linear and Quadratic Time-Frequency Signal Representations,"IEEE Signal Processing Mag.,Apr.1992,pp.21-67.

[2] T.A.C.M. Classen and W.F.G. Mecklenbrauker, "The Aliasing Problem in Discrete-Time Wigner Distributions," IEEE Trans. Acoust., Speech, Sig. Proc., pp.1067-1072, Oct. 1983.

[3] W. Martin and P. Flandrin, "Wigner-Ville analysis of nonstationary processes," IEEE Trans. Acoust., Speech, Signal Proc., vol. ASSP-33, pp. 1461-1470, Dec. 1985.

[4] F. Peyrin and R. Prost, "A Unified Definitions for the Discrete-Time, Discrete-Frequency and Discrete-Time/Frequency Wigner Distributions," IEEE Trans. Acoust., Speech, Sig. Proc., pp.858-867, Aug. 1986.

[5] L. Cohen, "Time-Frequency Distributions - A Review," Proc. IEEE, vol. 77, no. 7, pp. 941-981, July 1989.

[6] D. L. Jones and T. W. Parks, "A Resolution Comparison of Several Time-Frequency Representations," IEEE Trans. Sig. Proc., vol. 40, no. 2, pp. 413-420, Feb. 1992.

Chapter 2

[1] D.W. Baker," Pulsed Doppler Blood Flow Sensing,"IEEE Trans. Sonics Ulreason.,vol. SU-17,1970,pp.170-185.

[2] M. Brandestini," Topoflow-A digital full-range Doppler velocity meter," IEEE Trans. Sonics Ultrason.,vol. SU-25, no.5,1978,pp.287-293.

[3] W.D. Barber, J.W. Eberhard, S.G. Carr, "A new time domain technique for velocity maesurements using Doppler ultrasound," IEEE Trans. Biomed. Eng., vol. BME-32,Mar. 1985,pp.213-229.

[4] C. Kasai, K. Namekawa, A. Koyano, R. Omoto, "Real-time two dimensional blood flow imaging using an autocorrelation technique," IEEE Trans. Sonics Ultrason., vol. SU-32, May 1985,pp.458-464.

[5] N. Aydin, D.H. Evans, "Implementation of directional Doppler techniques using a digital signal processor," Med.& Biol.Eng.& Comput., 1994, vol.32, pp.157-164.

[6] J.A. Jensen, "Artifacts in blood velocity estimation using ultrasound and cross-correlation," Med.& Biol.Eng.& Comput., 1994, vol.32, pp.165-170.

[7] X. Zhang, G.H. Harrison, E.K. Balcer-Kubiczek, " Exposimetry of Unfocused Pulsed Ultrasound," IEEE Trans. Ultra., Ferroel., Freq. Contr., vol.41, no.1, Jan. 1994,pp.80-83.

[8] M.A. Benkhelifa, M. Gindre, J. Le Huerou, W. Urbach, " Echography Using Correlation Techniques: Choice of Coding Signal," IEEE Trans. Ultra., Ferroel., Freq. Contr., vol.41, no.5, Sep. 1994,pp.579-587.

[9] V.Tagliasco, G. Valli, "Eidetic bioengineering. Computational methods for medical images." CNR, Pàtron Editore, 1988.

[10] B. Angelsen, " Istantaneous frequency, mean frequency, and variance of mean frequency estimators for ultrasonic blood velocity Doppler signals," IEEE Trans. Biomed. Eng., vol. BME-28, no. 11, pp.733-741, 1981.

[11] A. Herment, G. Demoment, P. Dumée, J.P. Guglielmi, A. Delouche, " A New Adaptive Mean Frequency Estimator: Application to Constant Variance Color Flow Mapping," IEEE Trans. Ultra., Ferroel., Freq. Contr., vol.40, no.6,pp.796-804, Nov. 1993.

[12] R.I.Kitney and H. Talhami, " The zoom Wigner transform and its applications to analysis of blood velocity waveforms," J. Theor. Biol., vol. 129, pp.395-409, 1987.

[13] P. Leclerc, "Quantification of Noisy MRS Signals with the PWD," IEEE Trans. Biomed. Eng., vol. 41, no. 9, Sept. 1994.

Chapter 3

[1] B.Boashash and P.J.Black,"An Efficient Real-Time Implementation of the Wigner-Ville Distribution,"IEEE Trans. Acoust.,Speech,Signal Process.,vol.ASSP-35,Nov. 1987,pp.1611-1618.

[2] MATLAB, The MathWorks Inc., Natick, MA, 1992.

[3] W. Martin and P. Flandrin, "Wigner-Ville analysis of nonstationary processes," IEEE Trans. Acoust., Speech, Signal Proc., vol. ASSP-33, pp. 1461-1470, Dec. 1985.

Chapter 4

[1] S.V. Russell, D. McHugh, B.R. Moreman, "A programmable Doppler string test object," Phys. Med. Biol., vol. 38, 1993, pp.1623-1630.

[2] S.K. Holland, "Estimation of blood flow parameters using pulse Doppler ultrasound with corrections for spectral broadening," Ph.D. dissertation, Yale Univ., New Haven, CT, 1985.

[3] P. Rao, F.J. Taylor, "Estimation of istantaneous frequency using the discrete Wigner distribution," Electronics Letters, vol. 26, no.4, 15 th Feb., pp.246-248, 1990.

[4] K.M. Wong, Q. Jin, "Estimation of the Time-Varying Frequency of a Signal : The Cramer-Rao Bound and the Application of Wigner Distribution," IEEE Trans. Acoust., Speech, Signal Processing, vol. 38, no.3, March, 1990, pp.519-536.

[5] B. Boashash, "Estimating and interpreting the istantaneous frequency of a signal.-Part 2 : Algorithms and applications," Proc. IEEE, vol. 80, pp.540-568, Apr. 1992.

[6] G. Andria, M. Savino, A. Trotta, "Application of Wigner-Ville Distribution to Measurements on Transient Signals," IEEE Trans. Instrumen. Meas., vol. 43, no. 2, pp. 187-193, Apr., 1994.

[7] T.A.C.M. Classen and W.F.G. Mecklenbrauker, "The Wigner distribution-A tool for time-frequency signal analysis, Part 1- Continuous signals," Philips J. Res., vol. 35, no. 3, pp.217-250, 1980.

[8] A.Zeira, E.M.Zeira and S.K.Holland, "Pseudo-Wigner Distribution for Analysis of Pulsed Doppler Ultrasound," IEEE Trans., Ultrason., Ferroelec., Freq. Contr., vol.41, NO.3, May 1994, pp.346-352.

www.ingramcontent.com/pod-product-compliance
Lightning Source LLC
Chambersburg PA
CBHW081053170526
45165CB00006B/2261